USMLE ROAD MAP

BIOCHEMISTRY

LANGE

USMLE ROAD MAP

BIOCHEMISTRY

RICHARD G. MACDONALD

Department of Biochemistry and Molecular Biology
University of Nebraska Medical Center
Omaha, Nebraska

WILLIAM G. CHANEY

Department of Biochemistry and Molecular Biology
University of Nebraska Medical Center
Omaha, Nebraska

New York   Chicago   San Francisco   Lisbon   London   Madrid   Mexico City
Milan   New Delhi   San Juan   Seoul   Singapore   Sydney   Toronto

The McGraw·Hill Companies

USMLE Road Map: Biochemistry

1 2 3 4 5 6 7 8 9 0 DOC/DOC 0 9 8 7

ISBN-13: 978-0-07-144205-3
ISBN-10: 0-07-144205-7

Notice

Medicine is an ever-changing science. As new research and clinical experience broaden our knowledge, changes in treatment and drug therapy are required. The authors and the publisher of this work have checked with sources believed to be reliable in their efforts to provide information that is complete and generally in accord with the standards accepted at the time of publication. However, in view of the possibility of human error or changes in medical sciences, neither the authors nor the publisher nor any other party who has been involved in the preparation or publication of this work warrants that the information contained herein is in every respect accurate or complete, and they disclaim all responsiblity for any errors or omissions or for the results obtained from use of the information contained in this work. Readers are encouraged to confirm the information contained herein with other sources. For example and in particular, readers are advised to check the product information sheet included in the package of each drug they plan to administer to be certain that the information contained in this work is accurate and that changes have not been made in the recommended dose or in the contraindications for administration. This recommendation is of particular importance in connection with new or infrequently used drugs.

This book was set in Adobe Garamond by Pine Tree Composition, Inc.
The editors were Jason Malley, Harriet Lebowitz, and Penny Linskey.
The production supervisor was Sherri Souffrance.
The illustration manager was Charissa Baker.
Project Management was provided by Pine Tree Composition, Inc.
Graphics and illustrations created by Dragonfly Media Group.
The index was prepared by Ann Salinger.

RR Donnelley was the printer and binder.
This book is printed on acid-free paper.

International edition: ISBN-13: 978-0-07-110476-0; ISBN-10: 0-07-110476-3
Exclusive rights by The McGraw-Hill Companies, Inc., for manufacture and export. This book cannot be re-exported from the country to which it is consigned by McGraw-Hill. The International Edition is not available in North America.

CONTENTS

USING THE
USMLE ROAD MAP SERIES
FOR SUCCESSFUL REVIEW

What is the Road Map Series?

Short of having your own personal tutor, the *USMLE Road Map* Series is the best source for efficient review of major concepts and information in the medical sciences.

Why Do You Need A Road Map?

It allows you to navigate quickly and easily through your biochemistry and genetics course notes and textbook and prepares you for USMLE and course examinations.

How Does the Road Map Series Work?

Outline Form: Connects the facts in a conceptual framework so that you understand the ideas and retain the information.

Color and Boldface: Highlights words and phrases that trigger quick retrieval of concepts and facts.

Clear Explanations: Are fine-tuned by years of student interaction. The material is written by authors selected for their excellence in teaching and their experience in preparing students for board examinations.

Illustrations: Provide the vivid impressions that facilitate comprehension and recall.

 Clinical Correlations: Link all topics to their clinical applications, promoting fuller understanding and memory retention.

 Clinical Problems: Give you valuable practice for the clinical vignette-based USMLE questions.

 Explanations of Answers: Are learning tools that allow you to pinpoint your strengths and weaknesses.

COMMON ABBREVIATIONS

ADP	adenosine diphosphate
AMP	adenosine monophosphate
ATP	adenosine triphosphate
CNS	central nervous system
FAD	flavin adenine dinucleotide (oxidized form)
FADH$_2$	flavin adenine dinucleotide (reduced form)
GDP	guanosine diphosphate
GMP	guanosine monophosphate
GTP	guanosine triphosphate
HDL	high-density lipoprotein
LDL	low-density lipoprotein
NAD$^+$	nicotinamide adenine dinucleotide (oxidized form)
NADH	nicotinamide adenine dinucleotide (reduced form)
NADP$^+$	nicotinamide adenine dinucleotide phosphate (oxidized form)
NADPH	nicotinamide adenine dinucleotide phosphate (reduced form)
P$_i$	inorganic orthophosphate
PP$_i$	inorganic pyrophosphate
RBC	red blood cell
VLDL	very low-density lipoprotein
WBC	white blood cell

ACKNOWLEDGMENTS

The authors wish to thank all those listed in the credits for their assistance in the assembly of this book. In addition, we thank Janet Foltin, Harriet Lebowitz, Jennifer Bernstein, and our anonymous scientific editors for all that they have taught us in this process.

CHAPTER 1
PHYSIOLOGIC CHEMISTRY

I. Water

  A. The special chemical properties of **water** make it ideal as the main physiologic solvent for polar substances in the body.

    1. Within the water molecule, the oxygen nucleus draws electrons away from the hydrogen atoms, producing an internal charge separation that makes each molecule magnetic or **polar.**

    2. Substances that dissolve well in water are referred to as polar or **hydrophilic.**

    3. Molecules that dissolve sparingly in water are nonpolar or **hydrophobic.**

  B. Water molecules bind with each other through important **noncovalent interactions** called hydrogen bonds.

    1. Hydrogen bonds result from attraction between the partially positively charged hydrogen atoms of one molecule and the electronegative atom, usually oxygen or nitrogen, of another molecule.

    2. Hydrogen bonds are **weak** and rapidly break and re-form up to 10^{12} times per second in water at 25°C.

  C. The hydrogen bond network of water molecules confers special properties on water that are important for sustaining life.

    1. Water has a high **surface tension** where it comes in contact with air.

      a. Surface tension is the force acting to push together the liquid molecules at an air-liquid interface.

      b. This property causes the liquid to form droplets and to resist passage of substances across the interface.

      c. The surface tension of fluid at the **alveolar air-water interface** of the lungs contributes to **elastic recoil** that causes the alveoli to return to the original volume after **inflation** during breathing.

    2. Water has a high **heat of vaporization,** ie, the amount of heat needed to convert from liquid to gas phase. In conjunction with its **high heat capacity,** this property allows water to carry away heat efficiently as it evaporates, which accounts for the cooling effects of **perspiration.**

    3. Water has a high **dielectric constant,** which is a measure of its ability to carry electrical current, as it does in nerve cells.

II. Electrolytes

  A. **Electrolytes** are compounds that separate or **dissociate** in water into a positively charged **cation** and a negatively charged **anion.**

I

B. Because of their polar nature, electrolytes are **soluble** in water.
   1. The dissolved ions become surrounded by water and so have little tendency to re-associate at low concentrations.
   2. Important cationic electrolytes in human physiology include Na^+, K^+, Ca^{2+}, and Mg^{2+}, whereas Cl^- and HCO_3^- are critical anionic electrolytes.

III. Acids and Bases

A. Molecules that act as **proton donors** are **acids,** while those that act as **proton acceptors** are **bases.**
   1. Strong acids, such as hydrochloric acid (HCl), and strong bases, such as sodium hydroxide (NaOH), dissociate completely when dissolved in water.
   2. Most acids of physiologic importance are **weak acids,** which tend to dissociate reversibly into a proton and a conjugate base.

$$HA \rightleftarrows H^+ + A^-$$

   a. Physiologically important weak acids include **carboxylic acids** (such as acetic, carbonic, citric, and lactic acids), **phosphate**-based compounds, and **sulfated** molecules.
   b. In solutions of weak acids, an **equilibrium** is established between the **undissociated acid, HA,** and its **conjugate base A^-** that is defined by the equilibrium constant for dissociation of the acid, K_a.

$$K_a = \frac{[H^+][A^-]}{[HA]}$$

   c. The relative strengths of weak acids can be compared by converting their K_a values to **pK_a,** whose units correspond directly with the pH scale; the lower the value of an acid's pK_a, the greater the tendency for protons to dissociate.

$$pK_a = -\log K_a$$

   3. Weak bases that are important in physiology include **ammonia** and all compounds that have **amino ($-NH_3^+$)** groups, eg, amino acids and sugar amines.
   a. Dissociation of a weak base, BH^+ in the equation below, is also described by an equilibrium equation.

$$BH^+ \rightleftarrows B + H^+$$

   b. Many weak bases have pK_a values above 7.0, which reflect the tendency to retain rather than give up their proton.

IV. pH

A. Water is a weak acid that dissociates into a **proton, H^+,** and a **hydroxide ion, OH^-.**

$$H_2O \rightleftarrows H^+ + OH^-$$

   1. This **dissociation** is reversible and is defined by the **equilibrium constant, K_{eq}.**

$$K_{eq} = \frac{[H^+][OH^-]}{[H_2O]}$$

2. In pure water, very few water molecules actually undergo this dissociation, and the concentration of water is considered to be a constant, equal to 55.5 M.

3. Reorganization of the equilibrium equation gives a new, combined constant, K_w, and the **ion product of water.**

$$K_w = K_{eq} \times 55.5\ M = [H^+][OH^-]$$

4. Thus, in pure water, the $[H^+] = [OH^-] = 10^{-7}\ M$, and when acid or base is added to water, these ions change concentration in a reciprocal manner.

B. The acid state of a solution is represented by its **pH,** which is calculated from the $[H^+]$.

$$pH = -\log[H^+]$$

1. In pure water, where $[H^+] = 10^{-7}\ M$, the **pH = 7.0;** at this pH, the solution is considered **neutral.**

2. When the pH is < 7.0, the solution is **acidic;** when the pH is > 7.0, the solution is **basic** or **alkaline.**

3. Human plasma has a pH of 7.4 under normal conditions.
   a. Maintenance of plasma pH within a narrow range, 7.35 to 7.45, supports the optimal activity of enzymes and function of proteins.
   b. Deviation of plasma pH from this physiologic range interferes with the function of enzymes and proteins and, therefore, of cells.

4. In contrast, gastric fluid is more acidic (pH = 1.2–2.8) and pancreatic secretion is more alkaline (7.8–8.5).

DRUG ABSORPTION IN THE DIGESTIVE TRACT DEPENDS ON PH

CLINICAL CORRELATION

- *Ionized* or charged forms of drugs that are weak acids or bases cannot cross biologic membranes readily because of the nonpolar nature of the lipids that form the membrane bilayer.
- In the acidic environment of the stomach, drugs that are **weak acids,** such as aspirin, are in their protonated or **nonionized** form, which can be taken up by the **gastric mucosal cells.**
- Amine-based drugs, such as **oral antihistamines,** are **weak bases** that are absorbed well by the mucosal cells lining the **small intestine,** where the pH is alkaline and the drugs tend to lose their protons and become nonionized.

V. Buffers

A. Solutions of weak acids and bases act as **buffers** that resist changes in pH when acid or base is added (Figure 1–1).

B. The **Henderson-Hasselbalch equation** is derived from the rearrangement of the equilibrium equation for dissociation of a weak acid.

$$pH = pK_a + \log \frac{[\text{conjugate base}]}{[\text{conjugate acid}]} = pK_a + \log \frac{[A^-]}{[HA]}$$

1. The Henderson-Hasselbalch equation describes the relationship between the pH, the pK_a, and the concentrations of the conjugate acid and base.

2. The effectiveness of a buffering system is maximal when it is operating at a pH near its pK_a (Figure 1–1).
   a. When pH $\cong$ pK_a, the buffer is poised to absorb either added H^+ or OH^- with minimal change in pH.

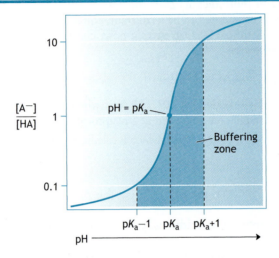

Figure 1-1. Weak acids act as buffers in a pH range near their pK_as. According to the Henderson-Hasselbalch equation, when the ratio of conjugate base to conjugate acid, $[A^-]/[HA]$ is plotted versus pH, a titration curve is generated that indicates a region of good buffering at pH = $pK_a \pm 1$ pH unit.

b. **Buffering capacity** is also related to the **buffer concentration.** For example, the ability of a weak acid solution to buffer added acid is related to the concentration of conjugate base available to combine with the protons.

C. The **carbonic acid-bicarbonate system** is the most important buffer of the blood.

1. **Carbonic acid, H_2CO_3,** is a weak acid that dissociates into a proton and the **bicarbonate anion, HCO_3^-** (Figure 1–2).
2. The carbonic acid-bicarbonate buffer system has a pK_a of 6.1, yet is still a very effective buffer at pH 7.4 because it is an **open buffer system,** in which one component, CO_2, can equilibrate between blood and air.

$$CO_2 + H_2O \rightleftarrows H_2CO_3 \rightleftarrows H^+ + HCO_3^-$$

a. This system is very **flexible** in response to changes in pH of the blood or the peripheral tissues.
b. **Dissolved CO_2** is in **equilibrium with gaseous CO_2** in the alveoli, which allows the lungs to help maintain blood pH by adjusting the amount of CO_2 expired.
c. An increase in CO_2 expiration shifts the carbonic acid-bicarbonate equation to the left (decreasing $[H^+]$); a decrease shifts it to the right (increasing $[H^+]$).
3. Dissolved CO_2 can combine with water to form carbonic acid, so CO_2 may be considered an acid from the physiologic standpoint.
4. Bicarbonate ion concentration is regulated mainly by excretion and synthesis in the **kidneys.**

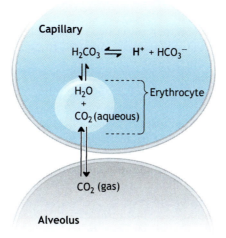

Capillary

$$H_2CO_3 \rightleftharpoons H^+ + HCO_3^-$$

$$H_2O$$
$$+$$
$$CO_2 \text{ (aqueous)}$$

Erythrocyte

$$CO_2 \text{ (gas)}$$

Alveolus

Figure 1–2. The carbonic acid-bicarbonate buffer system of the blood is responsive to alterations in P_{CO_2} within the alveoli by diffusion between the gas and aqueous phases.

METABOLIC ACIDOSIS

- *Alterations in metabolism that produce excess acid can cause blood pH to drop below 7.35, causing a metabolic* **acidosis.**
- *Examples of conditions that can lead to production of excess acid include diabetic ketoacidosis, lactic acidosis, sepsis, and renal failure.*
- *Excess acid is partially managed by* **respiratory compensation,** *by which increased depth and speed of expiration* **(hyperventilation)** *of CO_2 helps expel some of the acid, in addition to increased H^+ excretion in the urine.*
- *In the most serious cases or in the absence of treatment, metabolic acidosis may lead to unconsciousness, coma, or death.*

METABOLIC ALKALOSIS

- *Metabolic* **alkalosis** *may occur because of a loss of H^+ or due to retention of excess HCO_3^-, which may result from the following:*
  –Loss of stomach acid through excessive vomiting.
  –Ingestion of an alkalinizing drug such as sodium bicarbonate.
  –Changes in renal HCO_3^- balance in response to aldosterone or treatment with diuretics.
- *Excess HCO_3^- is managed to some extent by respiratory compensation* **(hypoventilation)** *but mainly by an increase in renal HCO_3^- excretion.*
- *If the pH remains above 7.55, as in severe alkalosis, arteriolar constriction may lead to reduced cerebral blood flow, tetany, seizure or, potentially, death.*

VI. Amphipathic Molecules

A. Substances that have both a hydrophilic group and a hydrophobic region, often a hydrocarbon tail, are referred to as **amphipathic.**

B. Amphipathic molecules do not dissolve fully in water but instead cluster together to form specialized structures with their polar groups oriented toward the water and nonpolar regions pointed away from the water.

1. **Micelles** are spherical structures that have the polar groups on the outside surface where they form hydrogen bonds with water, and the nonpolar tails are clustered in the core of the structure.

2. An important structure formed by amphipathic molecules is the lipid **bilayer,** in which the hydrocarbon tails line up in a parallel array with the hydrophilic head groups facing the polar fluids on either side.

3. **Lung surfactant** is a mixture of proteins and **amphipathic lipids** that acts like a detergent or soap to greatly decrease the surface tension forces at the alveolar fluid-air interface.

   a. The main surfactant protein apoprotein SP-A mingles with water molecules to interfere with the hydrogen bond network near the surface.

   b. The lipid components have their polar head groups inserted into the alveolar fluid and hydrophobic tails oriented toward the air.

LUNG SURFACTANT AND RESPIRATORY DISTRESS SYNDROME

- *The effect of **surfactant** to reduce the surface tension of the fluid lining the **alveoli** contributes to **decreased elastic recoil** and thereby **increases compliance** of the lung.*
- *Surfactant synthesis is stimulated immediately before birth in response to a surge of maternal **corticosteroid.***
- *Up to 15% of premature infants and even some babies delivered by cesarean section have inadequate levels of surfactant, producing **respiratory distress syndrome,** which is characterized by cyanosis and symptoms of labored breathing.*
- *Treatment options include corticosteroid administration to the mother prior to a cesarean section to induce surfactant production, direct tracheal **instillation of surfactant,** and in the most severe cases, mechanical ventilation.*

CLINICAL PROBLEMS

1. The weak organic acid, lactic acid, has a pK_a of 3.86. During strenuous exercise, lactic acid can accumulate in muscle cells to produce fatigue. If the ratio of the conjugate base form lactate to the conjugate acid form of lactic acid in muscle cells is approximately 100 to 1, what would be the pH in the muscle cells?

   A. 1.86

   B. 2.86

   C. 3.86

   D. 4.86

   E. 5.86

2. A patient arrives in the trauma center suffering from unknown internal injuries as a result of a traffic accident. She is semiconscious with a blood pressure of 64/40 mm Hg and appears to be going into shock. Blood gases reveal a P_{CO_2} of 39 mm Hg (normal = 40 mm Hg) and a bicarbonate of 15 mM (normal = 22–30 mM), with pH = 7.22. The best course of action to manage this patient's acidosis would be to start intravenous administration of a solution of:

   A. Sodium bicarbonate

   B. 5% dextrose

   C. Sodium lactate

   D. Sodium hydroxide

   E. Normal saline

3. Infants born prematurely are at risk for respiratory distress syndrome. In such cases, it is common to administer surfactant, the purpose of which is to alter which of the following properties of water at the alveolar interface with air?

   A. Surface tension

   B. Evaporation

   C. Heat of vaporization

   D. Ionization

   E. Dielectric constant

4. Lactic acid is considered to be a weak acid because:

   A. It is insoluble in water at standard temperature and pressure.

   B. It fails to obey the Henderson-Hasselbalch equation.

   C. Little of the acid form remains after it dissolves in water.

   D. The equilibrium between the acid and its conjugate base has a pK_a of 5.2.

   E. The lactate anion has minimal tendency to attract a proton.

5. The composite pK_a of the bicarbonate system, 6.1, may appear to make it ill-suited for buffering blood at physiologic pH of 7.4. Nevertheless, the system is very effective at buffering against additions of noncarbonic acids. Changes in the bicarbonate/carbonic acid ratio in such cases can be regulated by:

   A. Recruitment of bicarbonate reserves from the peripheral tissues.

   B. Conversion of carbonic acid to CO_2 and excretion in the urine.

   C. Conversion of carbonic acid to CO_2 followed by removal by the lungs.

   D. Reaction of excess carbonic acid with the amino termini of blood proteins.

   E. Binding of carbonic acid by hydroxide ions from the fluid phase of blood.

ANSWERS

1. **The answer is A.** The ratio of conjugate base to its acid for a physiologic buffer helps determine the pH of a solution according to the terms of the Henderson-Hasselbalch equation. When the concentration of base equals that of the acid form, the ratio is 1.0 and the pH = pK_a. In this case, a ratio of acid to base of 100:1 inverts to a base to acid ratio of 1:100 and calculates pH = 1.86. Such a highly acidic condition is never actually achieved within muscle cells because other weak acids, including those provided by inorganic phosphates and proteins, help buffer the solution by binding excess protons arising from dissociation of the lactic acid.

2. **The answer is C.** The normal P_{CO_2} value coupled with a low bicarbonate value and pH of 7.32 indicates a metabolic acidosis due to shock arising from the trauma. This condition can be managed by administration of a solution of the conjugate base of a weak acid. Although it may seem that sodium bicarbonate would be the natural choice to rapidly increase blood pH and replenish bicarbonate, this treatment should be reserved for severe cases of acidosis because of its risk of kidney damage. The best treatment option is to administer sodium lactate, which helps replace fluid loss due to potential internal bleeding as well as buffer some of the acid. Sodium gluconate solution would be an alternative option. Both of these agents help buffer the acid and are better tolerated by the kidneys than bicarbonate. Sodium hydroxide is a strong base and highly toxic. Dextrose (glucose) would not affect blood pH in this case. Normal saline would be valuable for fluid replenishment but has no buffering capability.

3. **The answer is A.** Lung surfactant reduces surface tension of the fluid lining the alveoli to increase pulmonary compliance and facilitate exchange of gases dissolved in that fluid from inspired air into the airway epithelial cells and eventually by diffusion into the blood. Although all the other options represent properties of water or solutions, they have nothing to do with the properties of surfactant.

4. **The answer is D.** Weak acids like lactic acid never completely dissociate in solution and are thus defined by the property that at least some of the protonated (undissociated acid) form and the unprotonated (conjugate base) form of the acid are present at all concentrations and pH conditions. The indicated pK_a of 5.2 is consistent with the idea that the lactate anion retains a strong affinity for protons, a hallmark of a weak acid. The lactate anion is highly water-soluble. All weak acids obey the Henderson-Hasselbalch equation.

5. **The answer is C.** Ingestion of an acid or excess production by the body, such as in diabetic ketoacidosis, may induce metabolic acidosis, a condition in which both pH and HCO_3^- become depressed. In response to this condition, the carbonic acid-bicarbonate system is capable of disposing of the excess acid in the form of CO_2. The equilibrium between bicarbonate and carbonic acid shifts toward formation of carbonic acid, which is converted to CO_2 and H_2O in the RBC catalyzed by carbonic anhydrase, an enzyme found mainly in the RBC. The excess CO_2 is then expired by the lungs as a result of respiratory compensation for the acidosis (Figure 1–2). The main role of the kidneys in managing acidosis is through excretion of H^+ rather than CO_2.

CHAPTER 2
PROTEIN STRUCTURE AND FUNCTION

I. Amino Acids

A. The amino acids are the **building blocks of proteins.**

  1. The 20 **amino acids** that cells use to make proteins have a common core structure.

   a. Most amino acids have a central carbon atom to which is attached a hydrogen atom, an **amino group, NH_3^+,** and a **carboxyl group, COO^-.**

   b. The **side chain** or R group distinguishes each amino acid chemically.

  2. Assembly of the amino acids to form **peptides** and proteins occurs by stepwise fusion of the carboxyl group of one amino acid with the amino group of another, with loss of a molecule of water during the reaction to form a **peptide bond.**

  3. Proteins can have a broad diversity of structures depending on their **amino acid sequences** and **composition.**

  4. The central carbon and the atoms involved in end-to-end linkage of the amino acids form the **polypeptide backbone,** with the side chains protruding outwardly to **interact** with other parts of the protein or with other molecules.

B. The 20 common amino acids can be classified into groups with similar side chain chemistry.

  1. The **nonpolar** or hydrophobic amino acids—glycine, alanine, valine, leucine, and isoleucine–have **alkyl** side chains (or simply a hydrogen atom in the case of glycine).

  2. Serine and threonine are small, polar amino acids that have **hydroxyl groups.**

  3. The **sulfur-containing** amino acids are cysteine and methionine.

  4. The **aromatic** amino acids, phenylalanine, tyrosine, and tryptophan, have ring structures and are nonpolar with the exception of the hydroxyl group of tyrosine.

  5. The **acidic** amino acids, aspartic acid and glutamic acid, have carboxyl groups.

  6. The **amides** of the carboxylic amino acids, asparagine and glutamine, are uncharged and polar.

  7. Members of the **basic** group, histidine, lysine, and arginine, have weak-base side chains.

  8. **Proline** is unique; it is an **imino acid** because its side chain loops back to form a five-membered **ring** with its amino group, which causes proline to produce **kinks** in the polypeptide backbone.

II. Charge Characteristics of Amino Acids and Proteins

A. The ionic properties of proteins at pH 7.4 are determined by the mixture of their acidic and basic amino acids.

    1. The carboxyl groups of **acidic amino acids,** aspartic acid and glutamic acid, have pK_a values < 5.0.

        a. These groups are thus unprotonated at neutral pH and contribute a **negative charge.**

        b. When these amino acids are in their unprotonated states, they are referred to as **aspartate and glutamate.**

    2. The **carboxyl-terminal** end of most proteins has a pK_a of 2.5–4.5 and thus is negatively charged at neutral pH.

    3. The side chains of the **basic amino acids** tend to retain their protons at neutral pH, and thereby contribute a **positive charge.**

        a. The imidazole ring of histidine has a pK_a of 6.5–7.5.

        b. The amino group of lysine exhibits a pK_a of 9.0–10.5.

        c. The guanidino group of arginine has a pK_a of 11.5–12.5.

    4. The **amino-terminal** end of most proteins also contributes a **positive charge** at neutral pH, since its pK_a is about 8.0.

B. Although titration curves for proteins are complex because of their multiple acidic and basic groups, their behavior can be illustrated by titration of a simple amino acid such as alanine (Figure 2–1).

    1. Alanine has two dissociable groups: the carboxyl group with pK_a = 2.5 and the amino group with pK_a = 9.5. A **buffering zone** is evident near each group's pK_a

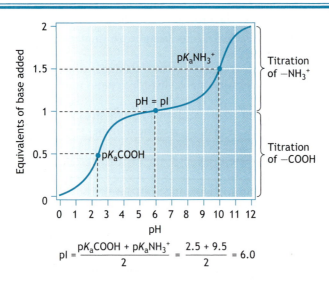

Figure 2–1. Titration of a solution of alanine with a strong base. One equivalent of base is the amount needed to titrate the protons from one group on all the alanine molecules present in the solution. Below the titration curve is a calculation of the pl for alanine derived as the mean of its two pK_a values.

as each of these groups releases its proton upon addition of a strong base (Figure 2–1).

2. At a pH where the protons from the carboxyl group have been completely removed but significant protons have not yet been released from the amino group, the charges on an amino acid balance, so the **overall charge is zero,** which defines the **zwitterion** state.

3. The pH at which an amino acid, a peptide, or a protein has zero overall charge after summing the contributions of all the charges is called the **isoelectric point (pI).**
    a. When **pH < pI,** the overall charge is **positive.**
    b. When **pH > pI,** the overall charge is **negative.**
    c. When **pH = pI,** there is **no overall charge.** A peptide or protein in such a case would not move in an electric field applied during electrophoresis.

III. Protein Structure

A. **Primary structure** refers to the linear **sequence** of amino acids linked by peptide bonds to make up a protein.

B. **Secondary structure** describes the **twisting** of the polypeptide **backbone** into regular structures that are stabilized by hydrogen bonding.
    1. The **α-helix** is a **coiled** structure stabilized by intrastrand hydrogen bonds (Figure 2–2).
        a. The structure is both **extensible and springy,** which contributes to the function of proteins that are primarily α-helix, such as **keratins** of fingernails, hair, and wool.
        b. Amino acid side chains project outward, away from the axis of the α-helix and decorate its exterior surface.
    2. **β-Sheet** structures are made from **highly extended** polypeptide chains that link together by hydrogen bonds between the neighboring strands and can be oriented in parallel or antiparallel arrays (Figure 2–2).
        a. Due to the very extended conformation of the polypeptide backbone, β-sheets resist stretching.
        b. The amino acid side chains project on either side of the plane of a β-sheet.
        c. Silk is composed of the protein **fibroin,** which is entirely β-sheet.

C. **Tertiary structure** is formed by combinations of secondary structural elements into a **three-dimensional organization** that is mainly stabilized by noncovalent interactions, such as hydrogen bonds.
    1. **Protein folding** is the complex process by which tertiary structures form within the cell.
    2. Regions of proteins that are capable of folding independently and that often have distinct functions are called **domains.**
    3. The side chains of highly **polar** amino acids tend to reside on the **exterior** of proteins, where they can form hydrogen bonds with water.
    4. The side chains of **nonpolar** amino acids are normally clustered in the **interior** of proteins to shield them from water.

D. **Quaternary structure** occurs in proteins that have multiple polypeptide chains, called **subunits.**
    1. In most cases, as in **hemoglobin,** the subunits are held together by noncovalent interactions.

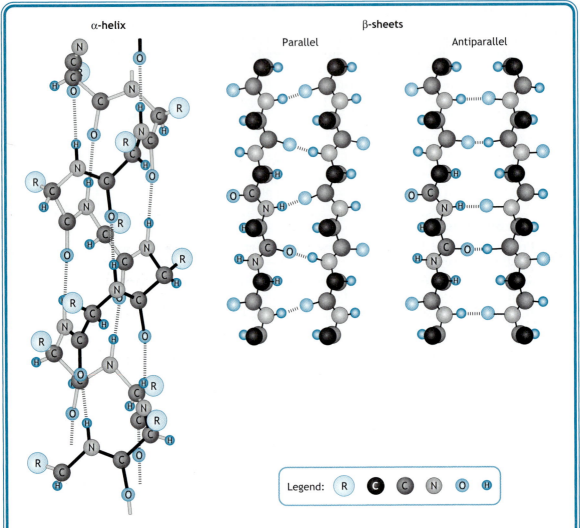

Figure 2–2. Structures of α-helix and β-sheet. Dashed lines indicate hydrogen bonds that stabilize these types of secondary structure. The hydrogen bonds of the α-helix are intrastrand, ie, formed between the backbone carbonyl oxygen and the amide hydrogen four amino acids up the helix. R groups represent the side chains in the α-helix. Side chains that would project above and below the plane of the page in the β-sheet structures have been omitted for clarity. Hydrogen bonds stabilizing the β-sheet are interstrand, ie, formed between groups on neighboring strands.

   2. In some multisubunit proteins, such as **immunoglobulins,** the subunits are held together by disulfide bonds or other covalent interactions.

CYSTIC FIBROSIS

• *Failure of a critical chloride transport protein to fold properly into its functional conformation contributes to many cases of cystic fibrosis (CF), which is the most common fatal inherited disorder of white people.*

- The gene responsible for CF codes for the **cystic fibrosis transmembrane conductance regulator (CFTR),** which is a **chloride channel** expressed on the surface of epithelial cells that line the affected organs.
- Approximately 70% of CFTR mutants worldwide are due to deletion of a single phenylalanine **(ΔF508)** that interferes with CFTR folding; this mutant CFTR is recognized as **abnormal** and is **degraded** (broken down).
- Patients with **CF** suffer from **thick mucous secretions** in the airways as well as the pancreas and intestinal lining due to impaired chloride absorption and consequent **fluid imbalance.**
  – These thick mucous secretions are difficult for the mucociliary cells of the airway to clear, resulting in chronic airway obstruction, **inflammation,** and **frequent lung infections.**
  – Decreased secretion of pancreatic enzymes leads to impairment of the digestive functions of the intestine.

IV. Collagen

A. Collagen is an abundant protein that provides the **structural framework** for tissues and organs.

    1. The long rod-like shape of collagen provides **rigidity and strength** to support the architecture of organs and tissues and to make **connective tissue.**

    2. Procollagen chains undergo extensive modification that strengthens the mature collagen molecules.

B. Collagen is composed of three highly extended chains that wrap around each other tightly in a **triple helix** (Figure 2–3).

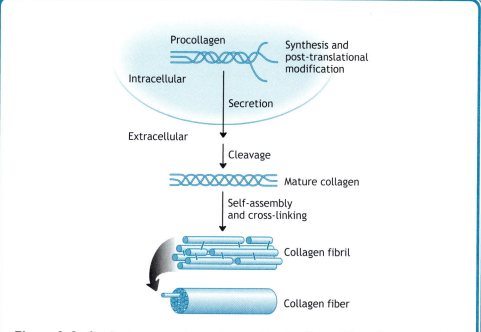

Figure 2–3. Synthesis, processing, and assembly of collagen. Note that many of the steps of final assembly that contribute to the strength of collagen fibers take place outside the cell.

1. Collagen is high in **glycine, proline,** and the modified amino acids **hydroxyproline** and hydroxylysine.

      2. Every third amino acid in most collagen chains is glycine, in **triplet repeats** of the sequence Gly-Pro-X and Gly-X-hydroxyproline, where X = any amino acid.

      3. The high frequency of glycine, with its small side chain, allows the three collagen chains to pack very tightly together for strength.

      4. Hydrogen bonding between the chains further stabilizes the triple helix.

   C. Much of collagen's strength arises from the special mechanism of its synthesis, post-translational modification, and assembly into **collagen fibers** (Figure 2–3).

      1. Covalent **cross-linking** of collagen chains adds markedly to the strength of the triple helix as well as to the larger structures formed by these connections.

         a. The first step in cross-linking is post-translational modification of some lysine residues in collagen to **allysine,** catalyzed by the enzyme **lysyl oxidase.**

         b. Allysine then reacts spontaneously with nearby lysine amino groups to form the cross-link.

      2. The final steps of collagen post-translational modification, including assembly of **collagen fibrils** and collagen fibers, occur after the protein has been secreted from the cell.

VITAMIN C DEFICIENCY

- Vitamin C, ascorbic acid, is required as a cofactor for the enzyme **prolyl hydroxylase,** which catalyzes the formation of hydroxyproline during collagen biosynthesis.
- **Vitamin C deficiency** leads to **impaired collagen production** and **defective collagen structure,** which causes weakening of the capillary walls and ultimately, of the **dentine** in teeth and the **osteoid** of bones.
- These biochemical defects are responsible for the pathophysiology of **scurvy,** characterized by generalized weakness, bleeding from the gums, loosening of the teeth, and formation of red spots surrounding hair follicles and underneath the fingernails from bleeding (**hemorrhage**).

EHLERS-DANLOS SYNDROME

- Defects in collagen synthesis, structure, or assembly into fibers are the principal basis for a group of connective tissue disorders called **Ehlers-Danlos syndrome (EDS).**
- There are many types of EDS, but they are generally characterized by **hyperextensible skin and joints,** poor wound healing and **"cigarette paper" scars** (ragged, gaping malformed scars), bruising, and other structural manifestations.
- At least 10 types of this heterogeneous group of disorders have been recognized, of which **type I** (gravis) is the most severe.
- Many types of EDS are inherited in an **autosomal dominant** manner because the mutant collagen chains interfere with function of the normal proteins with which they interact.

OSTEOGENESIS IMPERFECTA

- **Brittle bone disease,** or osteogenesis imperfecta (OI), is caused by mutations or absence of one of the genes encoding **type I collagen** chains, which interferes with assembly and function of the triple helix.
- OI is an inherited disorder characterized by a tendency to suffer multiple fractures because of **bone fragility,** due to poor formation of its collagen cement base.
- Four types of OI are distinguished clinically and differ in the types of genetic alterations that cause them as well as severity; the most severe form is **type II,** which is frequently lethal soon after birth.

• Other symptoms of OI include **blue sclerae,** bone deformities, short stature (types III and IV only), and hearing loss.

V. The Oxygen Binding Proteins—Myoglobin and Hemoglobin

A. **Myoglobin** is the primary **oxygen (O_2) storage** protein in **muscle,** where it binds O_2 with **high affinity.**

1. The heme group is held in a hydrophobic crevice of myoglobin and is made up of a porphyrin ring that forms four coordinate covalent bonds with the Fe^{2+} (ferrous iron) in its center.

2. In addition to interactions with the porphyrin ring, the heme Fe^{2+} is bonded to two histidine residues of the protein; when oxygen binds to the Fe^{2+}, it displaces the distal histidine.

3. O_2 remains bound until the Po_2 in muscle is very low (< 5 mm Hg), eg, during intensive exercise, which causes O_2 to **dissociate** so that it can be used in **aerobic metabolism.**

B. **Hemoglobin** in RBCs is responsible for O_2 **transport** from the lungs to the tissues for use in metabolism.

1. Hemoglobin binds O_2 at the high Po_2 (100 mm Hg) of the lung capillary beds and transports it to the peripheral tissues, where Po_2 is lower (~30 mm Hg) and O_2 dissociates from hemoglobin.

2. Adult hemoglobin (HbA) is a **heterotetramer** of **two α** and **two β** subunits, each of which has a protein component called **globin** that has a structure similar to myoglobin. Each subunit also has a heme group with a Fe^{2+} atom at its center (Figure 2–4).

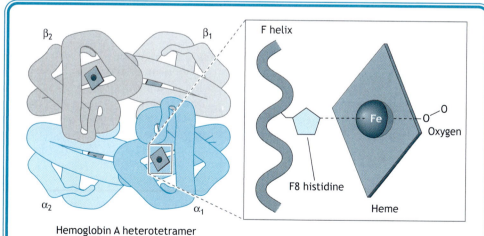

Figure 2–4. Structure of hemoglobin and its oxygen-binding site. An expanded view of the heme ring within the hydrophobic crevice is shown to the right. The polypeptide backbone of the nearby F helix is indicated by the ribbon with the imidazole ring of the F8 histidine residue projecting out as one of the ligands of the heme iron atom.

3. The hemoglobin heterotetramer is really a **dimer of dimers,** in which two αβ halves of the heterotetramer are held together at their interface by **noncovalent interactions.**
4. **Fetal hemoglobin (HbF),** which has slightly different O_2 binding properties from HbA, is composed of two α- and two γ-globin subunits.
   a. HbF has a higher affinity for O_2 at all P_{O_2} values than HbA, which facilitates **transplacental transfer** of O_2 from maternal blood to the fetal circulation.
   b. Switchover from expression of HbF to HbA occurs within 6 months of birth due to progressive shutdown of genes encoding the γ-globin chains and coordinate up-regulation of the genes for β-globin.

THALASSEMIAS

CLINICAL CORRELATION

- *Genetic defects that cause instability or reduced synthesis of either the α or β subunits of hemoglobin can cause **thalassemias,** which are characterized in most cases by **hemolytic anemia.***
- *The thalassemias are the most common disorders caused by mutations of a single gene worldwide; both **α-thalassemia** and **β-thalassemia** occur, depending on which subunit is deficient.*
- *Underproduction of β-globin chains in β-thalassemia leads to an excess of α chains, which can form an **$α_4$ tetramer** that precipitates in the RBCs as **inclusion bodies.***
- *The thalassemias are a diverse group of diseases with variable severity; patients are usually **anemic** and may have multiple organ manifestations due to excessive RBC death and tissue **hypoxia** (O_2 deficiency).*
- *The severity of β-thalassemia is reduced to a variable extent by the **persistence of HbF** production, which allows for continued presence of HbF in adult RBCs.*
- ***Incidence** of both thalassemias is high in **northern and central Africa,** the **Mediterranean** region, and across **southern Asia,** with a very high prevalence of α-thalassemia in Southeast Asia.*
- *Many inherited blood diseases show this geographic distribution, possibly because the altered RBC physiology **confers resistance to the malaria parasite,** which infects normal, HbA-bearing RBCs.*

5. There is no counterpart to the distal histidine of myoglobin in hemoglobin.
   a. The Fe^{2+}, which prefers six ligands, is coordinately bonded in hemoglobin at four positions by the porphyrin ring and in a fifth position by one histidine from the protein, with the sixth position being unfilled until O_2 binds.
   b. The five-liganded condition of the Fe^{2+} in hemoglobin distorts its structure and is important in initiating the conformational change that occurs on O_2 binding.
6. The O_2 saturation curve of hemoglobin is different from that of myoglobin (Figure 2–5), with increasing **affinity** of hemoglobin for O_2 as O_2 loading increases, indicating **cooperativity** of O_2 binding.
7. Hemoglobin alternates between two structurally and functionally distinct forms to fulfill its physiologic role.
   a. **Deoxyhemoglobin,** in which all four O_2 binding sites are unoccupied and which is also called the "T" or "taut" form, has **low O_2 affinity.**
   b. **Oxyhemoglobin,** to which four O_2 molecules are bound and which is also called the "R" or "relaxed" form, has **high O_2 affinity.**
8. Although the structure of deoxyhemoglobin resists loading of O_2, this resistance is overcome in the lungs by high P_{O_2}.

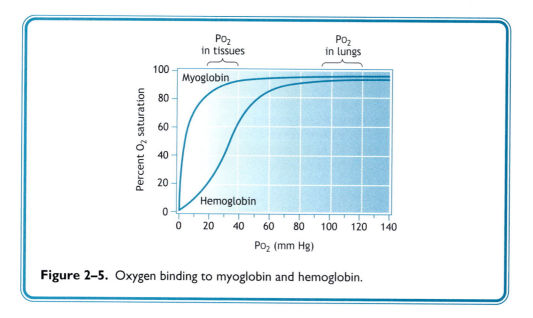

Figure 2–5. Oxygen binding to myoglobin and hemoglobin.

a. Binding of O_2 to the heme Fe^{2+} of one of the subunits causes a **conformational change** in the protein near the heme group that results from altered orientation of the Fe^{2+} in the plane of the porphyrin ring and a corresponding shift of the nearby protein structure.

b. This small shift is propagated through the protein backbone to force reorganization of noncovalent interactions at the dimer interface; some hydrogen bonds and **salt bridges** break and new ones characteristic of oxyhemoglobin are made.

c. In this way, the changes in structure of the subunit to which O_2 is bound are transmitted to the other subunits, each of which increases its affinity for and then binds O_2.

METHEMOGLOBINEMIA: OXIDATION OF HEME IRON

- **Methemoglobin** is a form of hemoglobin in which the iron atom is in the more **oxidized ferric (Fe^{3+})** state rather than the normal ferrous (Fe^{2+}) state.
  - Formation of methemoglobin occurs occasionally when O_2 carries away an electron as it dissociates from the heme iron.
  - Methemoglobin is not capable of binding oxygen, so it is normally reduced back to its functional state by an enzyme-mediated mechanism in the RBC.
- **Hereditary methemoglobinemia** arises from a deficiency of the enzyme that catalyzes this reduction, **NADH-cytochrome b_5 reductase**.
- This is a benign condition that causes patients to appear **cyanotic** and have **mild symptoms,** such as headache and fatigue.
- **Acquired methemoglobinemia** may occur in response to oxidizing agents, such as **sulfanilamide drugs, acetaminophen, benzocaine,** and **sodium nitroprusside,** which oxidize hemoglobin to methemoglobin, producing **cyanosis.**

- *Toxicity can be overcome by giving **methylene blue,** a dye that is metabolized to a form that reduces the Fe^{3+} of methemoglobin back to the Fe^{2+} state.*

9. Conditions in the peripheral tissues that stabilize the structure of deoxyhemoglobin promote dissociation of O_2.

   a. **CO_2** arising from metabolism must be carried back to the lungs for respiration and 10–15% is transported by covalent attachment to the amino-terminal ends of some of the hemoglobin subunits.

   b. The majority of CO_2 combines with water in a reaction catalyzed by **carbonic anhydrase** to form **carbonic acid** (see Chapter 1), which dissociates to bicarbonate and a proton, which is taken up by amino groups on hemoglobin.

   c. By altering noncovalent interactions between the $\alpha\beta$ dimers, both of the above effects favor conversion of hemoglobin from the oxy form to the deoxy form and, in so doing, enhance dissociation of O_2 from oxyhemoglobin in the tissues.

   d. The **Bohr effect** is the tendency of hemoglobin to release O_2 in response to **decreased pH,** conditions that prevail in metabolically active tissues.

      (1) **Binding of protons** to critical groups on hemoglobin **stabilizes deoxyhemoglobin** and thereby decreases the O_2 binding affinity of hemoglobin.

      (2) Conversely, increasing the pH promotes dissociation of protons from these groups on hemoglobin and favors return to the high-affinity state.

SICKLE CELL ANEMIA

CLINICAL CORRELATION

- *Sickle cell anemia is caused by synthesis of a mutant form of hemoglobin, **hemoglobin S ($\alpha_2\beta_{s2}$ or HbS),** in which a glutamic acid at position 6 of the hemoglobin β subunit is replaced by valine.*

- ***HbS has reduced solubility** in its deoxy form and tends to aggregate and distort the structure of RBCs, forming the characteristic **sickle cells** that clog small capillaries and cause **vasoocclusive crises.***

- *Patients with sickle cell anemia suffer fatigue and pain, which is frequently localized to the extremities, upon exertion or after exercise.*

- *HbS in RBCs confers **resistance to malaria** and thus the HbS allele occurs in highest frequency in people of African descent and is most prevalent in **West Africa.***

10. A byproduct of glycolysis, **2,3-bisphosphoglycerate (BPG)** is present in the RBCs at nearly equal concentration to that of hemoglobin, and it is a key regulator of O_2 affinity.

    a. **BPG binds** by making salt bridges with several positively charged residues in the **hemoglobin central cavity;** this cavity is large enough to accommodate BPG in deoxyhemoglobin but is too small for BPG to fit in oxyhemoglobin.

    b. BPG binding drives the oxy-to-deoxy conversion of hemoglobin and so promotes O_2 dissociation to facilitate delivery of O_2 to the tissues, where the Po_2 is low.

    c. In the lungs, Po_2 is high enough to force loading of O_2 to nearly saturate hemoglobin even in the presence of BPG.

    d. HbF does not bind BPG, which gives HbF a higher affinity for O_2 than HbA.

BPG RESPONSE TO HIGH ALTITUDE OR HYPOXEMIC CONDITIONS

- *Decreased P_{O_2} at **high altitude** leads to **reduced O_2 saturation** of hemoglobin as blood leaves the lungs.*
- ***BPG levels** are elevated in the RBCs of persons who have adapted to high altitude conditions, enhancing dissociation of O_2 in tissues to compensate for reduced O_2 saturation of hemoglobin.*
- *In conditions that lead to **chronic hypoxemia**, such as smoking and chronic obstructive pulmonary disease, an increased concentration of BPG in the RBCs promotes O_2 dissociation from hemoglobin in tissues to support cellular function.*

VI. Antibodies

A. **Antibodies** or **immunoglobulins (Ig)** are produced by **B lymphoid cells** in response to the presence of foreign molecules, usually proteins, nucleic acids, or carbohydrates, which are called **antigens.**
   1. Most antibodies have a complex quaternary structure, being composed of four individual polypeptide chains, **two heavy (H) chains and two light (L) chains.**
   2. The polypeptide chains are held together by **disulfide bonds** between the H and L chains within each half-molecule and between the H chains that join at the **hinge region.**
B. **Diversity** in the abilities of antibodies to recognize various antigens arises from differences in primary structure in the antigen-binding or **variable region.**
   1. The differences in sequence within the variable region produce a practically unlimited number of possible three-dimensional arrangements for the amino acid side chains to form the **complementarity-determining region (CDR),** which actually binds to the antigen.
   2. Antigen binding by the CDR occurs through **noncovalent interactions** that allow antibodies to be **specific** for structurally distinct antigens.
C. Antibodies are divided into five classes based on their **constant regions** and immune function.
   1. **IgM** molecules are the **first to appear** after antigen exposure and are unique in that they are made up of five antibody molecules coupled into a large array by disulfide bonding.
   2. **IgG** molecules are the most **abundant in plasma** and represent the main line of defense in the immune response.
   3. **IgA** molecules are secreted by and present in **mucous membranes** lining the intestine and the upper respiratory tract as well as in tears and the breast secretions milk and colostrum.
   4. The normal function of **IgD** molecules is not known.
   5. **IgE** molecules mediate the **allergic response.**

CLINICAL PROBLEMS

Some patients with erythrocytosis (excess RBCs) have a mutation that converts a lysine to alanine at amino acid 82 in the β subunit of hemoglobin. This particular lysine normally protrudes into the central cavity of deoxyhemoglobin, where it participates in binding 2,3-bisphosphoglycerate (BPG).

1. Which of the following effects would you predict this mutation to have on the affinity of hemoglobin for BPG and O_2, respectively, in such patients?

   A. Increase, Decrease

   B. Increase, Increase

   C. Decrease, Increase

   D. Decrease, Decrease

   E. No effect on either binding function

A 14-year-old girl is brought to the emergency department with shoulder pain and immobility consistent with dislocation. She is tall and thin and exhibits marked flexibility of her skin and joints—wrists, fingers, and ankles. There are no apparent cardiac abnormalities or vision problems. She has a past medical history of dislocation of both shoulders and her right hip, as well as easy bruising. Microscopic examination of a skin biopsy shows disorganized collagen fibers.

2. What is the most likely diagnosis in this case?

   A. Scurvy

   B. Osteogenesis imperfecta

   C. Prolyl hydroxylase deficiency

   D. Ehlers-Danlos syndrome

   E. Vitamin C deficiency

A 10-month-old white boy is being evaluated for weakness, pallor, hemorrhages under the fingernails, and bleeding gums. Radiographs indicate that bone near the growth plates shows reduced osteoid formation and grossly defective collagen structure.

3. What would be the most effective treatment for this patient's condition?

   A. Oral vitamin A

   B. Oral vitamin C

   C. Exclusion of dairy products from the diet

   D. Oral iron supplementation

   E. Growth hormone treatment

4. After first-time exposure to ragweed pollen, an initial immune response occurs followed by long-term sensitization to recurrent exposures to ragweed. Analysis for antibodies specific for the ragweed pollen would show immunoglobulins of which of the following classes at each stage of the immune response?

	Initial Exposure	Long-term Plasma Levels	Acute Allergic Response
A.	IgG	IgM	IgA
B.	IgD	IgG	IgA
C.	IgM	IgA	IgD
D.	IgG	IgA	IgE
E.	IgM	IgG	IgE

A 6-year-old black boy complains of acute abdominal pain that began after playing in a football game. He denies being tackled forcefully. He has a history of easy fatigue and several similar episodes of pain after exertion, with the pain usually restricted to his extremities.

5. Microscopic evaluation of his blood would be expected to reveal which of the following cellular abnormalities?

A. Increased WBC count

B. Deformed RBCs

C. Decreased WBC count

D. Increased RBC count (erythrocytosis)

E. Reduced platelet count

ANSWERS

1. The answer is C. Substitution of alanine for lysine removes from each β subunit a positive charge that is important for making a salt bridge with BPG. BPG should still bind but just not as well as it would to normal adult hemoglobin and the affinity would be decreased. Because BPG binding stabilizes the deoxy form of hemoglobin, reduced BPG binding affinity would make the deoxy-to-oxy transition occur at lower PO_2 values, ie, affinity of the mutant hemoglobin for O_2 would be increased.

2. The answer is D. Hyperextensibility of skin and hypermobility of joints are hallmark features of Ehlers-Danlos syndrome. The physical findings and history, especially the patient's tall, thin body, her joint and skin hyperextensibility and past medical history of dislocations, are consistent with a collagen disorder. Another inherited collagen disorder, osteogenesis imperfecta, is unlikely due to her tall stature and the absence of evidence of frequent fractures. Vitamin C deficiency affects collagen synthesis and structure but exhibits a different set of clinical findings (eg, hemorrhage).

3. The answer is B. The patient shows many signs of vitamin C deficiency or scurvy, which is seen most frequently in infants, the elderly, and in alcoholic patients. Particularly indicative of vitamin C deficiency are the multiple small hemorrhages that occur under the skin (petechiae) and nails and surrounding hair follicles. Bleeding gums are a classic indicator of scurvy.

4. The answer is E. Immune responses involving the soluble antibody or humoral system are initiated first in IgM class. Long-term immunity is mediated by IgG molecules that circulate in the plasma. Acute allergic responses frequently involve increased levels of IgE molecules.

5. The answer is B. Sickle cell anemia is caused by inheriting two copies of a mutant β globin gene that leads to synthesis of sickle hemoglobin, HbS. A severe case of sickle cell anemia would most likely have demonstrated symptoms and been diagnosed before the age of 6. However, he may only be a carrier, with one copy each of normal β-globin and one of the sickle allele, a condition called **sickle cell trait.** Nevertheless, the patient's symptoms are entirely consistent with an acute sickle cell crisis. These are brought on by exertion, which increases the levels of deoxyhemoglobin in RBCs. Under this condition, the mutant HbS molecules have reduced solubility; they tend to stick together in polymers that alter the shape of RBCs (sickle cells). Sickled RBCs are not as pliable as normal RBCs, so that they do not pass freely through the narrow passages of the capillaries and can cause clogging of microvessels. The pain experienced by this boy is likely due to such vasoocclusion in his joints and abdominal vessels.

CHAPTER 3
THE PHYSIOLOGIC ROLES OF ENZYMES

I. Enzyme-Catalyzed Reactions

A. Enzymes are catalysts that **increase the rate or velocity, *v*,** of many physiologic reactions.

   1. In the absence of enzymes, most reactions in the body would proceed so slowly that life would be impossible.

   2. Enzymes can **couple reactions** that would not occur spontaneously to an energy-releasing reaction, such as ATP hydrolysis, that makes the overall reaction favorable.

   3. Another of the most important properties of enzymes as **catalysts** is that they **are not changed** during the reactions they catalyze, which allows a single enzyme to catalyze a reaction many times.

B. Enzymes specifically bind the reactants in order to catalyze biologic reactions.

   1. During the reaction, the reactants or **substrates** are acted on by the enzyme to yield the **products.**

   2. Each substrate binds at its **binding site** on the enzyme, which may contain, be near to, or be the same as the **active site** harboring the amino acid side chains that participate directly in the reaction.

   3. Enzymes exhibit selectivity or **specificity,** a preference for catalyzing reactions with substrates having structures that interact properly with the catalytic residues of the active site.

C. A **deficiency in enzyme activity** can cause disease.

   1. Inherited absence or mutations in enzymes involved in critical metabolic pathways—eg, the urea cycle or glycogen metabolism—are referred to as **inborn errors of metabolism.** If not detected soon after birth, these conditions can lead to serious metabolic derangements in infants and even death.

   2. An enzyme deficiency can produce a **deficiency of the product** of the reaction it catalyzes, which may inhibit other reactions that depend on availability of that product.

   3. Accumulation of the substrate or metabolic byproducts of the substrate due to an enzyme deficiency can have profound physiologic consequences.

   4. Most inborn errors of metabolism manifest after birth because the exchange of metabolites between mother and fetus provides for fetal metabolic needs in utero.

   5. Therapeutic strategies for enzyme deficiency diseases include dietary modification and potential **gene therapy** or **direct enzyme replacement** (Table 3–1).

Table 3–1. Examples of enzyme replacement therapy for inherited diseases.

Disease	Enzyme Deficiency	Normal Function of the Enzyme	Major Symptoms or Findings on Examination	Physiologic Consequences and Prognosis
Pompe disease	Acid α-1,4-glucosidase	Hydrolysis of glycogen	Weakness, fatigue, failure to thrive, lethargy	Glycogen accumulation in several organs, including heart and skeletal muscle Congestive heart failure
Gaucher disease	Glucocerebrosidase	Hydrolysis of the glycolipid, gluco-cerebroside, a product of de-gradation of RBCs and WBCs	Easy bruising, fatigue, anemia, reduced platelet count	Accumulation of glucocerebroside in several organs, reduced lung and brain function, pain in upper trunk region, seizures, convulsions
Fabry disease	α-Galactosidase A	Hydrolysis of the lipid, globotria-osylceramide	Severe fatigue, painful paresthesias (numb-ness and tingling) of the feet and arms, purplish skin lesions on abdomen and buttocks	Accumulation of globotriaosylcer-amide in endothelial cells of the blood vessels, altered cellular structure of heart and glomeruli, renal failure

ALKAPTONURIA: DEFICIENCY OF HOMOGENTISATE OXIDASE

- *Homogentisate oxidase catalyzes an important reaction in **tyrosine metabolism,** which converts the **substrate homogentisic acid** to the **product maleylacetoacetic acid.***
- ***Inherited deficiency** of this enzyme in patients with **alkaptonuria** leads to accumulation of **ho-mogentisic acid,** which builds up in cartilage of the joints causing darkening of the tissue (**ochrono-sis**), inflammation, and **arthritis-like joint pain.***
- *Homogentisic acid is **excreted in urine,** which **darkens** when left standing exposed to oxygen.*

NIEMANN-PICK DISEASE: ACID SPHINGOMYELINASE DEFICIENCY

- ***Sphingomyelin,** a ubiquitous component of cell membranes, especially neuronal membranes, is nor-mally degraded within lysosomes by the enzyme **sphingomyelinase.***
- *In patients with **Niemann-Pick disease,** inherited deficiency of this enzyme causes spingomyelin to **accumulate in lysosomes of the brain,** bone marrow, and other organs.*
- *Enlargement of the lysosomes interferes with their normal function, leading to cell death and conse-quent **neuropathy.***

- Symptoms include **failure to thrive** and **death** in early childhood as well as **learning disorders** in those who survive the postnatal period.

HOMOCYSTINURIA: CYSTATHIONINE β-SYNTHASE DEFICIENCY

- **Cystathionine β-synthase** catalyzes conversion of homocysteine to cystathionine, a critical **precursor of cysteine.**
- Deficiency of this enzyme leads to the most common form of **homocystinuria,** a pediatric disorder characterized by **accumulation of homocysteine** and reduced activity of several sulfotransferase reactions that require this compound or its derivatives as substrate.
- **Accumulation of homocysteine and reduced transsulfation** of various compounds leads to abnormalities in connective tissue structures that cause altered blood vessel wall structure, loss of skeletal bone density (**osteoporosis**), **dislocated optic lens (ectopia lentis),** and increased risk of **blood clots.**

ENZYME REPLACEMENT THERAPY FOR INBORN ERRORS OF METABOLISM

- **Lysosomal enzyme deficiencies**, which frequently result in disease due to **accumulation of the substrate** for the missing enzyme, are suitable targets for **enzyme replacement therapy (ERT).**
- In ERT, **intravenously administered enzymes** are taken up directly by the affected cells through a receptor-mediated mechanism.
- ERT provides temporary relief of symptoms but must be given repeatedly and is not a permanent cure.

II. Enzyme Classification

- **A.** Enzymes can be made of either protein or RNA.
- **B.** Most enzymes are proteins, which are grouped according to the six **types of reactions** they catalyze (Table 3–2).
- **C.** Several important physiologic catalysts are made of RNA, and these RNA-based enzymes or **ribozymes** are of two general types.
  - **1.** RNA molecules that undergo **self-splicing,** in which an internal portion of the RNA molecule is removed while the parts on either side of this **intron** are reconnected (see Chapter 11).
  - **2.** Other RNA molecules that do not undergo self-splicing can act on other molecules as substrates are **true catalysts.**
    - **a.** Ribonuclease P cleaves transfer RNA precursors to their mature forms.
    - **b.** The 23S ribosomal RNA is responsible for the **peptidyl transferase** activity of the bacterial ribosome (see Chapter 12).
- **D. Isozymes** are protein-based enzymes that catalyze the same reaction but differ in amino acid composition.
  - **1.** Because of their structural differences, isozymes may often be distinguished by separation in an electric field (**electrophoresis**) or by reactivity with selective antibodies.
  - **2.** Several clinical uses have been made of isozymes selectively expressed by different tissues.

DIAGNOSIS OF HEART ATTACK AND MUSCLE DAMAGE

- The enzyme **creatine kinase (CK)** is formed of **two subunits** that can either be of the brain (B) type or the muscle (M) type, and different combinations of these types lead to isozymes that predominate in the **brain (BB), skeletal muscle (MM), and heart muscle (MB).**

Table 3–2. Classification of enzymes.

Class	Name	Trivial Names and Examples	Type of reaction catalyzed
1	Oxidoreductases	Dehydrogenases Reductases Oxidases	Addition or subtraction of electrons
2	Transferases	Kinases- Phosphotransferases Aminotransferases	Transfer of small groups: amino, acyl, phosphoryl, one-carbon, sugar
3	Hydrolases	Glycosidases Nucleases Peptidases	Add water across bonds to cleave them
4	Lyases	Decarboxylases Dehydratases Hydratases	Add the elements of water, ammonia, or carbon dioxide across a double bond (or the reverse reaction)
5	Isomerases	Mutases Epimerases	Structural rearrangements
6	Ligases	Synthases Synthetases	Join molecules together

- Within 3–4 hours of a **heart attack,** damaged myocardial cells **release CK of the MB type,** which can be detected in serum by a monoclonal antibody and is useful to confirm the diagnosis.
- Skeletal muscle **myopathy** often leads to release of CK of the MM type. **Rhabdomyolysis** is one of the major side effects of treatment with the cholesterol-lowering drugs the **statins.**
  – Inflammation of the muscle **(myositis)** leads to cell death.
  – The condition is characterized by muscle pain, weakness, **elevated CK MM,** and myoglobinuria.

III. Catalysis of Reactions by Enzymes at Physiologic Temperature

    A. The **rate** or **velocity** of any chemical reaction is measured as the change in concentration of reactants or products with time.

        1. Velocity decreases as reactants are used up to the point of **equilibrium,** where the overall **rate is zero.**

        2. The rates of most physiologic reactions depend only on the concentration of one reactant.

            a. Such reactions are said to obey **first-order kinetics.**

            b. Progress of such reactions can be followed according to the **half-life** of that reactant.

    B. The energy difference during conversion of reactants to products in a reaction can be represented by an **energy diagram** (Figure 3–1).

        1. The **activation energy** is the energy barrier that must be overcome to convert the reactants to products.

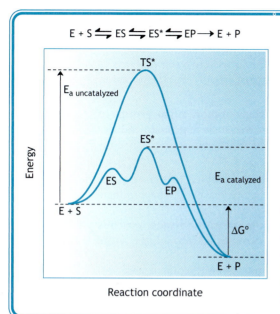

$$E + S \rightleftharpoons ES \rightleftharpoons ES^* \rightleftharpoons EP \longrightarrow E + P$$

Figure 3–1. Energy diagram for a reaction, comparing catalyzed and uncatalyzed conditions. The term $\Delta G°$ refers to the free energy change under standard conditions, ie, when reactants and products are present at 1 M concentrations.

2. As the reaction progresses and if sufficient activation energy is available, a state of high energy termed the **transition state** is reached; this state has a structure intermediate between reactants and products.

3. For a chemical reaction to occur **spontaneously,** the overall difference in **free energy (ΔG)** between products and reactants must be **negative** (Figure 3–1).

4. Like all catalysts, enzymes merely accelerate the rate but **do not change the ΔG** of a reaction or the equilibrium between reactants and products.

5. Enzymes reduce the activation energy of a reaction by providing an **alternative path** from reactants to products, one that may break up the reaction into smaller steps that are easier to overcome (Figure 3–1).

B. Many external factors other than catalysts can affect the rates of physiologic reactions.

1. In the absence of catalysis, a reaction can be accelerated by adding energy in the form of **heat,** but this is impractical in the body.

2. Increased **concentration** of one or more reactants also accelerates a reaction by increasing occupancy of substrate binding sites on available enzymes.

3. Enzymes normally operate within an **optimal pH range** in which the important amino acids of the active site have the correct state of protonation.

IV. Mechanisms of Enzyme Catalysis

A. Enzymes use a variety of strategies to catalyze reactions, and individual enzymes often use more than one strategy.

B. Substrate binding by an enzyme helps catalyze the reaction by bringing the reactants into **proximity** with the optimal **orientation** for reaction.

C. Amino acid side chains within active sites of many enzymes assist in catalysis by acting as **acids** or **bases** in reaction with the substrate.

1. In the mechanism of the pancreatic hydrolase **ribonuclease,** a specialized histidine within the active site acts as a **general acid or proton donor** to begin cleavage of the phosphodiester linkage of the substrate RNA.
2. The digestive enzyme **chymotrypsin** has a serine in its active site that acts as a **general base or proton acceptor** during hydrolysis of peptide bonds in protein substrates (Figure 3–2).

D. The binding of **polysaccharide** substrates that have six or more **sugar groups** to **lysozyme,** the enzyme in **tears** and **saliva** that cleaves such molecules, induces **strain** in the sugar nearest the active site making the nearby bond more susceptible to hydrolysis.

E. In **covalent catalysis,** the enzyme becomes covalently coupled to the substrate as an **intermediate** in the reaction mechanism before release of the products (Figure 3–2).
1. The active site serine of chymotrypsin attacks the protein substrate, which is cleaved and a portion of it becomes temporarily connected through the serine by an acyl linkage to the enzyme.
2. The acyl-enzyme intermediate reacts further by transfer of the polypeptide segment to water, completing cleavage (or **hydrolysis**) of the protein substrate.

SNAKE VENOM ENZYMES: HYDROLASES THAT PRODUCE TOXIC EFFECTS

- *Snake venoms are composed of a **toxic mixture of enzymes** that can kill or immobilize prey.*
- ***Neurotoxic** venoms of cobras, mambas, and coral snakes inhibit the enzyme **acetylcholinesterase.***
  - *This hydrolase normally breaks down the **neurotransmitter acetylcholine** within nerve synapses.*

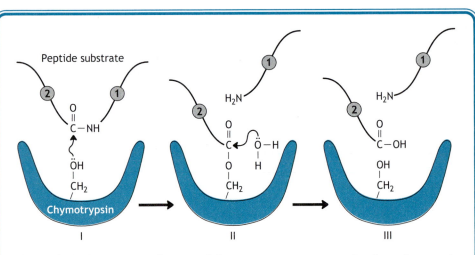

Figure 3–2. Reaction mechanism of chymotrypsin as an example of covalent catalysis. Step I involves attack of the enzyme's active site serine on the peptide bond to be cleaved. In step II, a covalent complex is formed between the enzyme and a portion of the substrate (peptide 2) with release of the rest of the substrate (peptide 1). Step III involves hydrolysis of the enzyme-substrate complex, which releases peptide 2 and completes the reaction.

– The resultant elevation of acetylcholine causes a transient period of contraction followed by **prolonged depolarization** in the postsynaptic muscle cell, which induces relaxation and then **paralysis** of the victim.

- **Hemotoxic** venoms of rattlesnakes and cottonmouths contain as their principal toxin **phosphodiesterase,** an enzyme that catalyzes hydrolysis of phosphodiester bonds in ATP and other substrates.
  – One consequence of this activity is altered metabolism of endothelial cells, which leads to cardiac effects and **rapid decrease in blood pressure.**
  –These venoms induce **circulatory shock** and potentially death.

ENZYMES AS THERAPEUTIC AGENTS

CLINICAL
CORRELATION

- The catalytic efficiency and exquisite specificity of enzymes have been exploited for use as therapeutic agents in certain diseases.
- Patients with **cystic fibrosis** use aerosol inhaler sprays of the DNA-hydrolyzing enzyme **deoxyribonuclease** to help reduce the viscosity of mucous secretions, which contain large amounts of DNA arising from destruction of WBCs as they fight lung infections.
- Patients who have had a **heart attack** or **stroke** are frequently treated by intravenous administration of **tissue plasminogen activator (tPA)** or **streptokinase,** enzymes that break down fibrin clots that clog blood vessels.

V. Kinetics of Enzyme-Catalyzed Reactions

    A. The rate of the simple enzyme-catalyzed reaction shown in the equation below can be described by **Michaelis-Menten kinetics.**

$$E + S \rightleftarrows ES \rightarrow E + P$$

    B. Most assays of enzyme activity depend on the assumption that very little of the **substrate, S,** has been converted into **product, P,** at the time of measurement.

        1. Under these **initial rate conditions,** the reaction described is being catalyzed only in the forward direction.

        2. The **velocity, v,** of the reaction depends on the substrate concentration up to a point when all the available enzymes are busy catalyzing the reaction at its maximal possible rate, V_{max} (Figure 3–3).

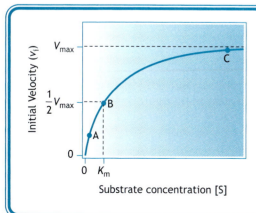

Figure 3–3. Relationship between [S] and v_i of an enzyme-catalyzed reaction.

C. The Michaelis-Menten equation describes the velocity, v, as a function of the substrate concentration, [S], for an enzyme-catalyzed reaction.

$$v_i = \frac{V_{max}[S]}{K_m + [S]}$$

1. Prominent in this equation is the term, K_m, defined as the substrate concentration, [S], at which the rate of the reaction is half-maximal, or $v = V_{max}/2$.
2. When [S] is well below K_m (Point A in Figure 3–3), then $[S] + K_m \cong K_m$, conditions where v is directly proportional to [S] and is low relative to V_{max}.
3. When $[S] = K_m$ (Point B in Figure 3–3), the Michaelis-Menten equation simplifies to $v = V_{max}/2$, which helps define the **physiologic range of [S]** at which the enzyme is best poised to respond to changing conditions of [S].
4. When [S] greatly exceeds K_m (Point C in Figure 3–3), $[S] + K_m \cong [S]$ and thus $v \cong V_{max}$ and the **enzyme is saturated.**
   a. Under this condition, adding more substrate to the reaction mixture does not further increase the rate.
   b. An example of this situation arises after a meal when the large influx of **glucose** into the liver **saturates hexokinase,** the low K_m enzyme responsible for its phosphorylation under low-glucose conditions.

ETHANOL SENSITIVITY DUE TO LACK OF A LOW-K_M ENZYME

- *Ethanol* is ordinarily metabolized in the liver by oxidation in two enzyme-catalyzed steps to *acetaldehyde* and ultimately *acetate.*
- *Some people exhibit **facial flushing** after consuming only modest amounts of **ethanol,** due to **acetaldehyde accumulation.***
- *Conversion of acetaldehyde to the less toxic acetate is catalyzed by one of several different types of **aldehyde dehydrogenase.***
- *Asians lack a form of aldehyde dehydrogenase with a **low K_m** for acetaldehyde and only express a high-K_m form of the enzyme, which allows increased blood levels of acetaldehyde sufficient to cause **vasodilation.***

D. Although it may seem from Point B in Figure 3–3 that the K_m can be determined from this representation of the velocity data, in practice, it is more accurate to use the **Lineweaver-Burk equation,** a modified form of the Michaelis-Menten equation, for estimation of K_m and V_{max} (Figure 3–4).

$$\frac{1}{v_i} = \left(\frac{K_m}{V_{max}}\right)\frac{1}{[S]} + \frac{1}{V_{max}}$$

1. Based on the Lineweaver-Burk equation, a plot of $1/v$ *versus* $1/[S]$ gives a straight line on a Lineweaver-Burk or **double reciprocal plot** (Figure 3–4).
2. V_{max} can be estimated from the y-intercept of this plot.
3. K_m can be derived by projecting the line back to the x-intercept.

VI. Enzyme Inhibitors

A. Enzyme inhibitors work in several ways and are clinically important as drugs.

B. **Competitive inhibitors** resemble the substrate in structure and **bind reversibly** to the enzyme's active site.

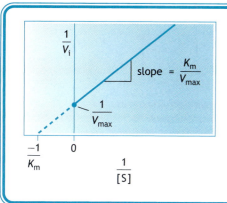

Figure 3–4. Lineweaver-Burk double-reciprocal plot of $1/v_i$ versus $1/[S]$ for estimation of K_m and V_{max} of an enzyme-catalyzed reaction.

   1. Because a competitive inhibitor binds to the same site on the enzyme as the substrate, it **can be displaced by increasing the substrate concentration,** which overcomes the inhibition.
   2. Competitive inhibitors **increase the apparent K_m** while having no effect on V_{max} (Figure 3–5).
C. **Noncompetitive inhibitors** bind to a site on the enzyme other than the substrate binding site to form an **inactive enzyme-inhibitor complex.**
   1. A noncompetitive inhibitor cannot be displaced from the enzyme by increasing substrate concentration.
   2. Noncompetitive inhibitors **decrease V_{max}** without effecting K_m (Figure 3–5).
D. **Irreversible inhibitors** are acted upon by the enzyme to form a **covalent complex** at the substrate binding site or active site of the enzyme.
   1. The covalent complex permanently inactivates the enzyme.
   2. Such inhibitors can only be used once, so they are often called **suicide inhibitors.**

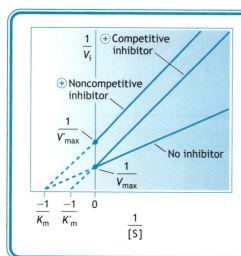

Figure 3–5. Lineweaver-Burk plots for inhibition of an enzyme-catalyzed reaction. K_m' and V_{max}' are the altered values representing the effect of the inhibitors.

3. A Lineweaver-Burk plot for an irreversible inhibitor resembles that of a non-competitive inhibitor.

ORGANOPHOSPHOROUS PESTICIDES: SUICIDE INHIBITORS OF ACETYLCHOLINESTERASE

- *Organophosphates form stable phosphoesters with the active site serine of **acetylcholinesterase,** the enzyme responsible for hydrolysis and inactivation of acetylcholine at cholinergic synapses.*
- ***Irreversible inhibition** of the enzyme leads to accumulation of acetylcholine at these synapses and consequent neurologic impairment.*
- ***Poisoning by pesticides** that contain organophosphate compounds produces a variety of symptoms, including nausea, blurred vision, fatigue, muscle weakness and, potentially, death caused by paralysis of respiratory muscles.*

MANY DRUGS ACT AS ENZYME INHIBITORS

- *Many drugs, including antibiotics and antiviral agents, operate by **inhibiting critical enzyme-catalyzed reactions** or serve as alternative **dead-end substrates** of such reactions.*
- *The antibiotic activity of **penicillin** is due to its ability to **inhibit transpeptidases** responsible for cross-link formation in construction of **bacterial cell walls,** leading to lysis of the weakened cells.*
- ***Sulfanilamides** are antibiotics that serve as **structural analogs of para-aminobenzoic acid (PABA),** a substrate in the formation of folic acid by many bacteria. Substitution of the sulfanilamide compound in place of PABA in the reaction **prevents formation of the critical coenzyme folic acid.***
- ***Inhibitors of the HIV protease** are useful in **antiviral therapy** strategies because this enzyme is absolutely required for processing of proteins needed for synthesis of the viral coat.*

VII. Coenzymes and Cofactors

    A. Coenzymes are **small organic molecules** that are required for activity of certain enzymes.

        1. Coenzymes **participate directly** in the enzyme-catalyzed reaction, often binding to one or more reactants.

        2. Some coenzymes bind loosely near the active site of the enzyme and thus act like substrates, while others are covalently bound to the enzyme as a **prosthetic group**.

        3. Many coenzymes are derived from **vitamins** (Table 3–3).

    B. Cofactors are **small inorganic ions** that are required for proper structure or to aid in catalysis for up to 70% of enzymes.

        1. Metalloenzymes have tightly bound metal ions, such as Zn^{2+} or Fe^{2+}, that serve as **metal ion bridges** between the enzyme and substrate.

        2. Some metal ions participate as **acids** to assist the enzyme in catalysis.

        3. Many metal ions can act as electron sinks, which allows them to participate in catalysis by **electron withdrawal** from the substrate, activating it toward reaction.

        4. In other cases, binding of a metal ion, such as Na^+, K^+, or Mn^{2+}, causes a **structural change** in the enzyme that is optimal for its activity.

        5. Metal ions in the form of **organometallic complexes** such as the iron atom in **heme** can undergo one-electron transfers in oxidation-reduction reactions catalyzed by **oxidoreductases** with associated **cytochromes.**

Table 3–3. Physiologic functions of coenzymes and cofactors.

Coenzyme/ Cofactor	Type of Binding	Derived from Vitamin	Physiologic Function
ATP	Loose	–	Phosphate donor in kinase reactions; energy donor in many reactions
NAD^+	Loose	Niacin (B_3)	Intermediate carrier of $2e^-$ and $2H^+$ in oxi-doreductase-catalyzed reactions
$NADP^+$	Loose	Niacin (B_3)	Same as NAD^+ but used mainly in biosynthetic pathways and detoxification reactions
FAD	Tight	Riboflavin (B_2)	Intermediate carrier of $2e^-$ and $2H^+$ in oxi-doreductase-catalyzed reactions
Flavin mononu-cleotide (FMN)	Tight	Riboflavin (B_2)	Same as FAD
Pyridoxal phosphate	Tight	Pyridoxine (B_6)	Intermediate carrier of amino groups during aminotransfer reactions
Thiamine pyrophosphate	Tight	Thiamine (B_1)	Cofactor for oxidative removal of CO_2 in several reactions of carbohydrate metabolism
Cobalamin compounds	Tight	Cobalamin (B_{12})	Transfer of methyl group to homocysteine during synthesis of methionine; metabolism of methylmalonyl coenzyme A
Tetrahydrofolic acid (THF)	Loose	Folic acid	Methyl group donor in one-carbon transfer reactions; critical in biosynthesis of purines and pyrimidines
Coenzyme A	Loose	Pantothenic acid (B_5)	Esterified to organic acids in many steps of fatty acid and carbohydrate metabolism
Biotin	Tight	Biotin	Intermediate carrier of CO_2 in carboxylation reactions
Ascorbic acid	Tight	Ascorbic acid (C)	Maintains reduced state of iron atom in enzymes involved in hydroxylation of proline and lysine in collagen

VIII. Allosteric Regulation of Enzymes

A. Key enzymes that catalyze **rate-limiting** steps of metabolic pathways or that are responsible for major cellular processes must be regulated to maintain **homeostasis** of individual cells and the organism overall.

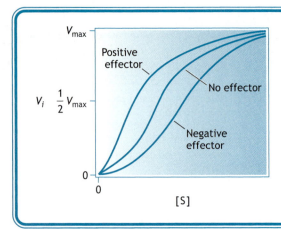

Figure 3–6. Relationship between v_i and [S] for a reaction catalyzed by an allosteric enzyme, showing the effects of positive and negative effectors.

B. **Allosteric regulation** refers to binding of a molecule to a site on the enzyme **other than the active site** and induces a subsequent **change in shape of the enzyme** causing an increase or decrease in its activity.

C. Many allosteric enzymes have **multiple subunits** whose interaction accounts for their unusual kinetic properties.
   1. Enzymes that are subject to allosteric regulation by either positive or negative **effectors** exhibit **cooperativity.**
   2. In the presence of **positive cooperativity,** a plot of v versus [S] shows **sigmoidal kinetics,** ie, is S-shaped (Figure 3–6).
      a. This kinetic behavior signifies that the enzyme's affinity for the substrate increases as a function of substrate loading.
      b. This is analogous to **O_2 binding by hemoglobin,** in which O_2 loading to one subunit facilitates O_2 binding to the next subunit, and so on.

D. **Feedback inhibition** occurs when the end product of a metabolic pathway accumulates, binds to and inhibits a critical enzyme upstream in the pathway, either as a competitive inhibitor or an allosteric effector.

CLINICAL PROBLEMS

A Polish man and his friend who is of Japanese descent are sharing conversation over drinks at a party. After the Polish man finishes his second bottle of beer, he notices that his friend, despite having drunk only half his drink, appears flushed in the face. His friend then complains of dizziness and headache and asks to be driven home.

1. The marked difference in tolerance to alcohol illustrated by these men is most likely due to a gene encoding which of the following enzymes?

   A. Alcohol dehydrogenase

   B. Acetate dehydrogenase

C. Alcohol reductase

    D. Aldehyde dehydrogenase

    E. Aldehyde aminotransferase

2. A noncompetitive enzyme inhibitor

    A. Decreases V_{max} and increases K_m.

    B. Decreases V_{max} and has no effect on K_m.

    C. Has no effect on V_{max} or K_m.

    D. Has no effect on V_{max} and increases K_m.

    E. Has no effect on V_{max} and decreases K_m.

A 47-year-old man is evaluated for a 12-hour history of nausea, vomiting and, more recent, difficulty breathing. His past medical history is unremarkable, and he takes no medications. However, he is a farmer who has had similar episodes in the past after working with agricultural chemicals in his fields. Just yesterday he reports applying diazinon, an organophosphate insecticide, to his sugar beet field.

3. After consultation with the poison center, you conclude that this patient's condition is most likely due to inhibition of which of the following enzymes?

    A. Acetate dehydrogenase

    B. Alanine aminotransferase

    C. Streptokinase

    D. Acetylcholinesterase

    E. Creatine kinase

Accidental ingestion of ethylene glycol, an ingredient of automotive antifreeze, is fairly common among children because of the liquid's pleasant color and sweet taste. Ethylene glycol itself is not very toxic, but it is metabolized by alcohol dehydrogenase to the toxic compounds glycolic acid, glyoxylic acid, and oxalic acid, which can produce acidosis and lead to renal failure and death. Treatment for suspected ethylene glycol poisoning is hemodialysis to remove the toxic metabolites and administration of a substance that reduces the metabolism of ethylene glycol by displacing it from the enzyme.

4. Which of the following compounds would be best suited for this therapy?

    A. Acetic acid

    B. Ethanol

    C. Aspirin

    D. Acetaldehyde

    E. Glucose

Glucose taken up by liver cells is rapidly phosphorylated to glucose 6-phosphate with ATP serving as the phosphate donor in the initial step of metabolism and assimilation of the sugar. Two enzymes, which may be considered isozymes, are capable of catalyzing this reaction in the liver cell. Hexokinase has a low K_m of ~0.05 mM for glucose, whereas glucokinase exhibits sigmoidal kinetics with an approximate K_m of ~5 mM. After a large meal, the glucose concentration in the hepatic portal vein may approximate 5 mM.

5. After such a large meal, which of the following scenarios describes the relative activity levels for these two enzymes?

	Hexokinase	**Glucokinase**
A.	Not active	Not active
B.	$v \cong \frac{1}{2}V_{max}$	Not active
C.	$v \cong V_{max}$	Not active
D.	$v \cong V_{max}$	$v \cong \frac{1}{2}V_{max}$
E.	$v \cong V_{max}$	$v \cong V_{max}$

ANSWERS

1. The answer is D. Many Asians lack a low-K_m form of acetaldehyde dehydrogenase, which is responsible for detoxifying acetaldehyde generated by oxidation of ethanol in the liver. Acetaldehyde accumulation in the blood of such individuals leads to the facial flushing and neurologic effects exhibited by the man of Japanese descent.

2. The answer is B. A noncompetitive inhibitor binds to the enzyme at a site other than the substrate binding site, so it has little measurable effect on the enzyme's affinity for substrate, as represented by the K_m. However, the inhibitor has the effect of decreasing the availability of active enzyme capable of catalyzing the reaction, which manifests itself as a decrease in V_{max}.

3. The answer is D. Organophosphates react with the active site serine residue of hydrolases such as acetylcholinesterase and form a stable phosphoester modification of that serine that inactivates the enzyme toward substrate. Inhibition of acetylcholinesterase causes overstimulation of the end organs regulated by those nerves. The symptoms manifested by this patient reflect such neurologic effects resulting from the inhalation or skin absorption of the pesticide diazinon.

4. The answer is B. The therapeutic rationale for ethylene glycol poisoning is to compete for the attention of alcohol dehydrogenase by providing a preferred substrate, ethanol, so that the enzyme is unavailable to catalyze oxidation of ethylene glycol to toxic metabolites. Ethanol will displace ethylene glycol by mass action for a limited time, during which hemodialysis is used to remove ethylene glycol and its toxic metabolites from the patient's bloodstream.

5. The answer is D. This problem provides a practical illustration of the use of the Michaelis-Menten equation. The high concentration of glucose in the hepatic portal vein after a meal would promote a high rate of glucose uptake into liver cells, necessitating rapid phosphorylation of the sugar. The glucose concentration far exceeds the K_m of hexokinase, ie, $[S] > K_m$, meaning that the enzyme will be nearly saturated with substrate and $v \cong V_{max}$. However, the $[S] \cong K_m$ for glucokinase, which will be active in catalyzing the phosphorylation reaction and $v \cong \frac{1}{2}V_{max}$.

CHAPTER 4
CELL MEMBRANES

I. Overview of Membrane Structure and Function

A. The main structural feature of biologic membranes is the **lipid bilayer** (Figure 4–1).

  1. The bilayer is composed of **amphipathic** lipid molecules oriented according to their preferences for interaction with water.

   a. Polar head groups face toward the aqueous environment of the intracellular and extracellular fluids.

   b. Nonpolar tails form a hydrophobic or fatty middle region of the bilayer.

  2. The major components of all biologic membranes are **lipids** and **proteins,** to which **sugars** may be attached.

B. Biologic membranes regulate the composition and the contents within the spaces they enclose.

  1. The **plasma membrane** enclosing the entire cell controls traffic of materials coming into and going out of the cell.

  2. The **organelles** are surrounded by membranes, which regulate the specialized functions within the assigned **compartments.**

II. Membrane Components: Lipids

A. The three major types of **amphipathic lipids** found in membranes are the glycerophospholipids (also called phosphoglycerides), the sphingolipids, and cholesterol.

  1. The glycerophospholipids and the phosphorylated derivatives of the sphingolipids are collectively called **phospholipids.**

  2. Phospholipids are responsible for organizing the bilayer structure of the membrane, whereas cholesterol's unique ringed structure allows it to regulate the fluidity of the membrane.

B. Glycerophospholipids have **two long-chain fatty acids** in an ester linkage to positions 1 and 2 of a **glycerol backbone** and a **phosphate** attached to position 3 (Figure 4–1).

  1. Members of the glycerophospholipid family are distinguished by the group attached via a phosphoester linkage to the phosphate of the polar head group.

   a. Many of these groups are bases, such as serine, ethanolamine, or choline.

   b. Cardiolipin is abundant in the inner mitochondrial membrane and is unusual because it is made up of two phosphatidic acids connected through a glycerol bridge.

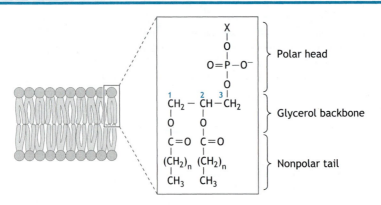

Figure 4–1. Structures of the membrane bilayer and an amphipathic phospholipid. The head group attachment, X, may be H as in phosphatidic acid or one of several substituents linked via phosphoesters in the glycerophospholipids. The nonpolar tail is depicted as composed of saturated fatty acids in this molecule. The overall length of the hydrocarbon chain of the fatty acids may vary from 14 to 20.

2. The **fatty acids** attached to the glycerol backbone also vary in length and structure (Figure 4–2).
    a. Fatty acids that have **no double bonds** between the carbons of their tails are thus **saturated** and form a straight hydrocarbon chain.
    b. Fatty acids that contain one or more **double bonds** are **unsaturated** because they have lost some electrons.
        (1) Most naturally occurring unsaturated fatty acids have *cis* **double bonds.**
        (2) The tail becomes fixed at each double bond, which reduces flexibility and causes the chain to **bend** at a 30-degree angle.

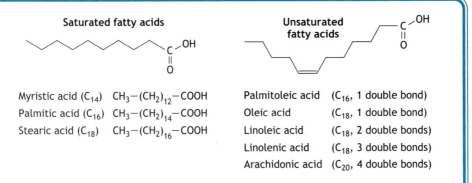

Figure 4–2. Structures of naturally occurring fatty acids. All the double bonds in these structures are of the *cis* configuration.

C. Sphingolipids are composed of a long-chain **fatty acid** connected to the **amino alcohols sphingosine** or **dihydrosphingosine.**

1. Attachment of another long-chain fatty acid in an amide linkage to the amino group of sphingosine forms a **ceramide,** the parent compound for many of the physiologically important sphingolipids.

2. Addition of a **phosphorylcholine group** to the ceramide converts the molecule into **sphingomyelin,** an important component of neuronal membranes.

3. By contrast, attachment of a **sugar** to the sphingosine forms a **glycosphingolipid,** which is also an important component of neuronal membranes, especially of the brain.

   a. **Glucose** and **galactose** are the main six-carbon sugars found in an important subclass of glycosphingolipids called the **cerebrosides,** forming glucocerebroside and galactocerebroside, respectively.

   b. The most complex glycosphingolipids are the **gangliosides,** which have an oligosaccharide structure containing **sialic acid** (eg, *N*-acetylneuraminic acid).

SCHINDLER DISEASE

- *Schindler disease (also called lysosomal α-N-acetylgalactosaminidase [**α-NAGA**] deficiency, Schindler Type) is 1 of the over 40 glycoprotein storage diseases.*
- *Deficiency or mutation of α-NAGA leads to an abnormal accumulation of some glycosphingolipids trapped in the lysosomes of many tissues of the body.*
- *Schindler disease type I, the classic form of the disease, begins in infancy.*
  - *This is a rare, metabolic disorder inherited in an autosomal recessive manner.*
  - *Children develop normally until 8–15 months of age, when they begin to lose previously acquired skills requiring coordination of physical and mental activities (**developmental regression**).*
  - *Other symptoms include decreased muscle tone (**hypotonia**) and weakness; involuntary, rapid eye movements (**nystagmus**); visual impairment; and seizures.*
- *Schindler disease type II, also known as **Kanzaki disease,** is an adult-onset form of the disease that causes milder symptoms that may not become apparent until the second or third decade of life.*
  - *Symptoms may include dilation of blood vessels over which clusters of wart-like discolorations grow on the skin (**angiokeratomas**).*
  - *Permanent widening of groups of blood vessels (**telangiectasia**) causing redness of the skin in affected areas is common.*
  - *Other symptoms include relative coarsening of facial features and mild cognitive impairment.*

D. Cholesterol is not only an important contributor to the structural properties of cell membranes, but it is also the precursor for **steroid hormone synthesis** and a major component of the **lipoproteins.**

1. Cholesterol has a **four-ringed structure** with a branched hydrocarbon chain attached to its 17 position and a polar hydroxyl group at position 3 (Figure 4–3).

2. The ring structure of cholesterol makes it **flat** and **very stiff.**

3. Consequently, its effect in the membrane is to increase the melting temperature or **decrease fluidity,** which has important effects on membrane functions, eg, transport and transmembrane signaling.

III. Organization of the Lipid Bilayer

A. Membranes are organized in the form of a two-dimensional array, as represented by the **fluid mosaic model.**

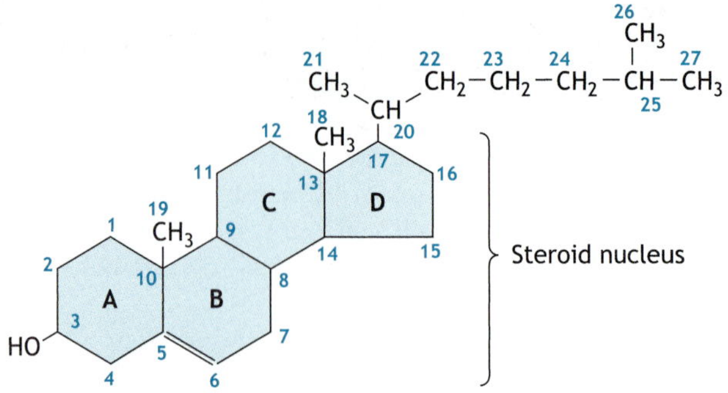

CH₃ group labeling: 26, 27, 25, 24, 23, 22, 21, 20, 18, 19, etc.

Figure: structure of cholesterol with labels

$$
\begin{array}{c}
\overset{26}{CH_3} \\
|
\end{array}
$$

$$\overset{21}{CH_3}\!-\!\overset{22}{CH_2}\!-\!\overset{23}{CH_2}\!-\!\overset{24}{CH_2}\!-\!\overset{}{\underset{25}{CH}}\!-\!\overset{27}{CH_3}$$

Steroid nucleus

Figure 4–3. Structure of cholesterol.

B. Proteins are embedded in, span across, or decorate the surfaces of the lipid bilayer.
   1. Integral membrane proteins are partially **embedded** in the hydrophobic center of the lipid bilayer.
      a. Protein regions that span the membrane must interact with the lipid zone and are thus **nonpolar.**
      b. If the protein has only a single **membrane-spanning (transmembrane) domain,** it is usually formed of an α-helix composed mainly of **nonpolar residues.**
      c. In contrast, if the protein has multiple transmembrane domains forming a channel, they will be oriented with polar amino acids facing the aqueous channel and nonpolar residues facing the lipids.
   2. Peripheral membrane proteins interact with the membrane loosely and often reversibly (Figure 4–4).
      a. Proteins may be bound by **charge-charge interactions** between charges on the surface of a membrane-embedded protein or the charges of the phospholipid head groups coating the membrane surface.
      b. In addition, proteins may interact with the lipid components of the membrane in several different ways (Figure 4–4).
C. Depending on the temperature and lipid composition, regions of the membrane may have different levels of **fluidity**—either fluid (partially liquid) or semicrystalline (partially solid).
   1. Membrane fluidity regulates **lateral movement** of proteins and lipids in the bilayer.
   2. Cholesterol tends to localize in the outer regions of the membrane, which makes the periphery less fluid than the center.
   3. Glycerophospholipids and cholesterol join together with specialized glycosyl phosphatidylinositol–linked proteins to form **lipid domains** or **rafts,** which move together as a unit laterally through the membrane.
   4. Unsaturated fatty acid chains do not pack together in the bilayer as tightly as saturated fatty acid chains; these properties contribute to different degrees of **fluidity** of membranes of different lipid composition.

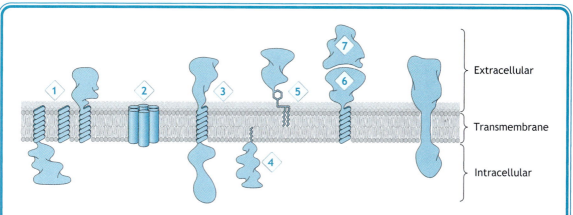

Figure 4–4. The domain organization of an integral, transmembrane protein as well as the mechanisms for interaction of proteins with membranes. The numbers illustrate the various ways by which proteins can associate with membranes: 1, multiple transmembrane domains formed of α-helices; 2, a pore-forming structure composed of multiple transmembrane domains; 3, a transmembrane protein with a single α-helical membrane-spanning domain; 4, a protein bound to the membrane by insertion into the bilayer of a covalently attached fatty acid (from the inside) or 5, a glycosyl phosphatidylinositol anchor (from the outside); 6, a protein composed only of an extracellular domain and a membrane-embedded nonpolar tail; 7, a peripheral membrane protein noncovalently bound to an integral membrane protein.

TRANS FATS AND ATHEROSCLEROSIS

- *The chemical process by which polyunsaturated vegetable oil is transformed to hard margarine or shortening produces fatty acids with* trans *as well as* cis *double bonds.*
- *During this hydrogenation process, the physical properties of the oils at room temperature are changed from liquid to solid.*
- *Unsaturated fats that have **trans double bonds** produced by hydrogenation and saturated fats with single bonds have similar linear hydrocarbon geometries, lipid packing properties, and effects on lipoprotein profiles of those who eat them.*
- *Many studies have now linked consumption of* trans *fats to elevated LDL or "bad" cholesterol levels, decreased HDL or "good" cholesterol levels, and a presumed higher risk of atherosclerosis, just as with saturated fats.*

ANESTHETIC AND ALCOHOL EFFECTS ON MEMBRANE FLUIDITY

- *Alterations in membrane fluidity, especially of **neurons**, can produce profound changes in cellular function.*
- *Anesthetics **increase membrane fluidity** due to their lipid solubility and ability to cause disordering of packed fatty acid tails in the bilayer, which is thought to interfere with the ability of neurons to conduct signals such as pain sensation to the brain.*
- *Although ethanol is **amphipathic,** it has substantial lipid solubility, and ethanol-induced intoxication and its ultimate anesthetic effect are also likely due to **increased fluidity** of neuronal membranes, resulting in **impairment of nerve conduction** to the CNS.*

IV. Membrane Components: Proteins

A. **Transmembrane proteins** have special structures that contribute to their specialized functions (Figure 4–4).
   1. The portion of the protein that protrudes above the plane of the membrane is the **extracellular domain.**
   2. The extracellular domain is linked to the **transmembrane domain,** which may be formed by up to 12 polypeptide strands that pass through the membrane.
   3. The portion of the protein that protrudes into the cytoplasm is the **intracellular domain,** which may be composed of a single folded section of polypeptide or by several loops and tails.

B. **Membrane proteins** have many different functions, which mainly relate to **intercellular communication** or **exchange of materials** with the environment.
   1. **Transporters** take up small molecules such as sugars, amino acids, and ions that otherwise cannot gain entry into the cell.
   2. **Receptors** mediate the actions of extracellular signals upon the cell (see Chapter 14).

C. Most membrane proteins undergo post-translational **glycosylation** to improve their interactions with the aqueous environment and to protect them from degradation by proteases.
   1. Sugars may be attached to **serine, threonine (O-linked),** or **asparagine (N-linked)** residues of the glycoproteins.
   2. The structures of oligosaccharides linked to these proteins can be complex and many of them contribute to **antigenicity,** the ability of the cell surface to elicit an immune response.

V. Membrane Components: Carbohydrates

A. Carbohydrates have a **carbon backbone bearing hydroxyl groups** with either an **aldehyde** or **ketone** at one carbon (Figure 4–5).

B. Simple sugars may take on several types of structures in solution.
   1. Simple sugars or **monosaccharides** are classified according to the number of carbons in the backbone.
      a. **Pentoses** have five carbons; examples include ribose and ribulose.
      b. **Hexoses** have six carbons: examples include glucose, galactose, fructose, and mannose.
   2. Most sugars are **asymmetric** and designated either D- or L- in stereochemistry.
   3. Simple sugars in aqueous solution usually form **cyclic structures,** either hemiacetals or hemiketals (Figure 4–5).
      a. The rings may have five or six members.
      b. Depending on how the cyclic structure was formed, the substituents at the connecting carbon may be **anomers**—having either α or β configuration.
      c. These forms of sugars are usually depicted by **Haworth projections.**
   4. The hexoses are structurally distinguished by **different configurations at one or more carbons.**
      a. **Diastereomers** are molecules differing in configuration at one or more carbons.
      b. **Epimers** are molecules that differ in their configurations at only one carbon, thus glucose and galactose are both epimers and diastereomers.

A

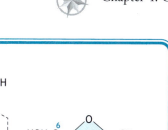

β-D-Glucose

β-D-Fructose

B

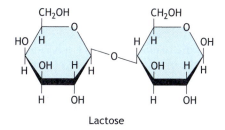

Sucrose

Lactose

C

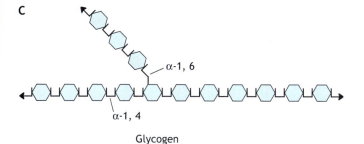

α-1, 6

α-1, 4

Glycogen

Figure 4–5. A: Cyclic structures of glucose and fructose. Glucose, an aldose, can form an intramolecular hemiacetal by reaction of the hydroxyl group on the fifth carbon (C-5) with the C-1 aldehyde. The six-membered ring formed in this way is called a pyranose. Fructose, a ketose, can undergo a similar intramolecular reaction between its C-5 hydroxyl and the C-2 keto group to form a five-membered fura-nose ring. The ring structures are shown as Haworth projections. **B:** Structures of sucrose and lactose. Sucrose, a nonreducing disaccharide, is composed of glucose and galactose connected by an α-1,2 linkage. Lactose, a reducing disaccharide, is formed of galactose connected to glucose by a β-1,4 linkage. **C:** Glycogen is the principal polysaccharide in human tissues and is made up of glucose molecules linked by α-1,4 bonds, with branches connected by α-1,6 linkage.

5. Modifications of one or more groups convert simple sugars into a variety of **sugar derivatives.**
   a. Replacement of −OH by −H converts the sugar into a **deoxymonosaccharide,** such as deoxyribose.
   b. Replacement of −OH by −NH$_2$ converts the sugar into an **amino sugar** designated as -**osamine,** eg, glucosamine.
   c. **Oxidation** of the terminal −CH$_2$OH to −COOH converts the sugar into a -**uronic acid,** such as glucuronic acid.

C. Sugars can be polymerized or interconnected to create chains termed **oligosaccharides** (≤ 8 sugars) or **polysaccharides** (> 8 sugars) (Figure 4–5).
   1. The linkage between sugars is formed by **condensation** of the hemiacetal or hemiketal of one sugar with a hydroxyl of another sugar with loss of water in the reaction.
   2. The linkage is called a **glycosidic bond** and can either be classified as α or β depending on the stereochemistry of the anomeric carbons at the bridge points.
   3. The important difference between α and β glycosidic bonds can be seen in the digestibility of the major plant polysaccharides cellulose and starch.
      a. **Cellulose,** the primary component of plant cell walls, is made up of **α–1,4-linked glucose,** which cannot be broken down by digestive enzymes. So humans cannot use cellulose as a direct dietary source of glucose.
      b. **Starch,** the main form of stored sugar in plants, is made up of **β–1,4-linked glucose,** which can be hydrolyzed by enzymes of the digestive tract, eg, α-amylase. Thus, starch is an important dietary source of glucose.

VI. Transmembrane Transport

A. **Polar molecules,** such as water, inorganic ions, and charged organic molecules, **cannot pass** unaided through the lipid bilayer of the membrane.
   1. Either a protein that acts as a **transporter** or that forms a **channel** or **pore** through the bilayer is needed to allow passage of such molecules.
   2. However, dissolved gases (such as O$_2$, CO$_2$, and N$_2$) can pass freely in either direction across membranes.

B. **Channels** allow passage of **small molecules** and **ions.**
   1. When open, a channel is a **water-lined pore** through which small, polar molecules can pass.
   2. Traffic through the channel is governed by **diffusion,** from higher concentration to lower.
   3. Channels do not bind the molecules that pass through them, but they **can be inhibited** or **regulated** by signals that cause the channel to open and close.
      a. Molecules pass very **rapidly** through open channels, at a rate of about 10^7 per second.
      b. Opening and closing of channels occur by **changes in conformation** of these integral membrane proteins.
      c. Some channels are regulated by **binding of an agonist neurotransmitter** (eg, acetylcholine regulation of the nicotinic-acetylcholine receptor, which is a Na$^+$ channel).
   4. Some channels are **voltage gated,** so that they open or close at a specific membrane potential to aid in **neurotransmission.**

a. In the neuron, **membrane depolarization** causes the Na^+ channel to open and allow the flow of Na^+ into the cell (an **inward Na^+ current**) during transmission of an electric impulse through the nerve.

b. There is a requirement for **insulation of the** neurons for proper transmission of the action potential through the gating of ion channels.

   (1) The **myelin sheath** forms by extension of the plasma membrane of neurons (Schwann cells) that wraps tightly many times around the extended cytoplasm.

   (2) The lipid nature of the myelin sheath makes it **water- and ion-impermeant,** and hence insulates the neuron to permit **transfer** or **propagation** of the electrical impulse.

KRABBE DISEASE

- As 1 of the 12 known *leukodystrophies,* Krabbe disease produces impaired myelin sheath development with *progressive neurodegeneration* of both the CNS and the peripheral nervous system.
  - Type I is the most severe form; patients are affected before 6 months of age and have a prognosis of death before age 2.
  - The onset of types II through IV may be delayed until late infancy through early adulthood.
- Children with Krabbe disease exhibit irritability, fever, seizures, limb stiffness, delayed mental or motor development, vomiting, feeding difficulties, hypertonia, spasticity, deafness, and blindness.
- The incidence of Krabbe disease is 1 in 100,000 births in the United States.
- Krabbe disease is caused by inherited deficiency of the *lysosomal hydrolase galactocerebrosidase,* the enzyme responsible for degradation of galactosylceramide, a component of the *myelin sheath,* and other galactosphingosines (eg, psychosine).
- Accumulation of psychosine is thought to cause toxicity and neuronal death.

C. **Transporters** within the membranes allow for **selective uptake** of specific molecules or classes of molecules and mediate two major types of transport—**passive** and **active.**

1. **Passive transport** or **facilitated diffusion** has no energy requirement and is defined as transport of molecules down their concentration gradient (high to low concentration).

2. **Active transport** is defined as transport against a concentration gradient and is accomplished by "pumps" that must be coupled to **energy expenditure** to make the process spontaneous.

   a. Many transporters that transport substances against a concentration gradient couple transport to ATP hydrolysis.

   b. Energy for transport may also be provided through simultaneous **dissipation of an ion** or **electrochemical gradient,** eg, glucose absorption by cells of the renal proximal tubule is coupled to simultaneous cotransport of Na^+ down its electrochemical gradient.

D. Transporters can be further distinguished according to the number and directions of the molecules they transport.

1. **Uniport** is when one substance is transported in a single direction, eg, the GLUT1 glucose transporter of the RBC.

2. **Cotransport** is when two or more molecules that move simultaneously or in sequence are transported.

a. Symport means substances are cotransported in the same direction.

    b. Antiport means substances are cotransported in opposite directions.

E. In contrast to channels, **transporters bind** and **assist** in movement of molecules as they cross the membrane and many of the steps involved are **analogous to the actions of enzymes** (Figure 4–6).

F. Transporters involved in **facilitated diffusion** are a diverse group, but they share the properties of **substrate specificity** and **saturability.**

  1. **Glucose transporters in muscle and fat tissue** operate by facilitated diffusion.

    a. The transporters are carriers that initiate their work by binding glucose on the outside of the membrane.

    b. The carrier undergoes a **conformational change** that exposes the bound glucose to the interior of the cell.

    c. Glucose released from the carrier is rapidly phosphorylated to **glucose 6-phosphate** by the enzymes hexokinase or glucokinase, which begins glucose metabolism (see Chapter 6).

    d. Glucose phosphorylation is so thorough that the intracellular concentration of free glucose in cells other than liver is effectively zero, meaning that the concentration gradient highly favors its uptake.

    e. Although ATP is the phosphate donor for glucose phosphorylation, ATP hydrolysis is not directly involved in glucose transport.

  2. The **chloride-bicarbonate exchanger** mediates **antiport** of the anions Cl^- and HCO_3^- in the membranes of renal tubule cells and the RBCs.

    a. The anions may move in either direction depending on the concentration gradients on either side of the membrane.

    b. The transporter is responsible for balancing bicarbonate ion concentrations in the RBC and for HCO_3^- **efflux** from the kidney to compensate for H^+ efflux.

G. Examples of **active transport** illustrate their range of mechanisms with the common theme of **energy requirement.**

  1. The **plasma membrane Na$^+$-K$^+$ ATPase** maintains intracellular Na$^+$ concentration low and intracellular K$^+$ concentration high relative to the extracellular fluid (Figure 4–7).

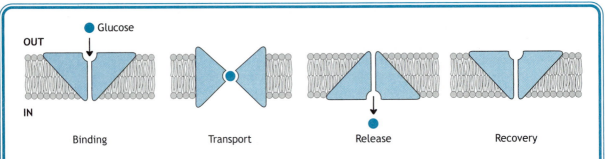

OUT

Glucose

IN

Binding   Transport   Release   Recovery

Figure 4–6. Mechanism of facilitated diffusion mediated by a glucose transporter. This is an example of uniport. The reversible interconversion between conformations of the transporter in which the glucose-binding site is alternately exposed to the exterior and interior of the cell is called a "ping-pong" mechanism.

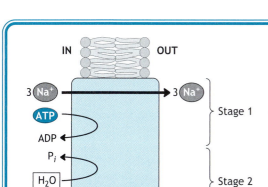

Figure 4–7. Schematic diagram of the plasma membrane Na⁺/K⁺ ATPase. The ATPase is an antiporter that operates in two stages. In the first stage, three Na⁺ are expelled from the cell, followed by a second stage during which two K⁺ are taken in. The reaction is catalyzed by ATP hydrolysis initiated during the first stage creating a phosphoenzyme intermediate that is hydrolyzed during the second stage to release orthophosphate (P_i).

    a. The ATPase is an **integral membrane pump** that exchanges three Na⁺ ions for two K⁺ ions.

    b. ATP is hydrolyzed to ADP + P_i via a catalytic site on the intracellular face of the protein.

    c. The action of the pump also serves to maintain a net negative electrical potential toward the inside of the cell.

  2. Amino acid uptake into epithelial cells of the intestinal lumen is mediated by **Na⁺/amino acid cotransporters.**

    a. This symport mechanism is specific only for the L-amino acids derived from digestion of dietary proteins.

    b. The **energy** for this concentrative mechanism of amino acid transport comes directly from the **Na⁺ electrochemical gradient across the brush border membrane.**

    c. There are **seven transport systems** tailored to chemically similar groups of amino acids, eg, there is one for neutral amino acids with small or polar side chains such as alanine, serine, and threonine.

HARTNUP DISORDER

- *Hartnup disorder is a rare condition caused by **impaired resorption of neutral amino acids** (especially tryptophan, alanine, threonine, glutamine, and histidine) in the renal tubules and malabsorption in the intestine, resulting from mutations that lead to defective function of a neutral amino acid transporter.*

- *Hartnup disorder exhibits **symptoms similar to pellagra** (niacin deficiency), characterized by three of the "four D's": diarrhea, dermatitis (a red, scaly rash), dementia (intermittent ataxia), and death (rarely).*

- *Patients show signs of tryptophan deficiency despite a healthy diet as well as elevated urinary and fecal excretion of the neutral amino acids.*

CYSTINURIA

- *Cystinuria, also called **cystine urolithiasis,** arises from **impaired reabsorptive transport of cystine** and the cationic amino acids from the fluid within the **renal proximal tubules.***
- *The biochemical defect is a deficiency or mutation of the gene that encodes the common membrane transporter for cystine and the dibasic amino acids.*
- *The disease is characterized by **excessive excretion of cystine and the dibasic amino acids** arginine, lysine, and ornithine by the kidneys that may lead to precipitation of some of these compounds in the form of **kidney stones.***
- *Symptoms of cystinuria, which develop during the teenage years to early adulthood, are those typically caused by recurring kidney stones, such as pain in the side or back often of a severe or debilitating nature.*
- *Cystinuria is an autosomal recessive disease with an incidence of 1 in 15,000 live births in the United States.*
- *The disease is classified into three subtypes, Rosenberg I, II, and III.*
  - *Type I is the most common variant caused by **mutation** or **deficient expression of a transporter.***
  - *Types II and III were thought to be allelic variants of this same transporter gene, but recent linkage analyses reveal type III to be a defect of a different transporter.*

CLINICAL PROBLEMS

A 21-year-old white woman arrives at the emergency department complaining of nausea, vomiting, and severe abdominal pain that have persisted for about 9 hours. She is doubled over in pain, even in the prone position. Physical examination reveals tenderness in the lower left abdomen and a mild fever. An abdominal radiograph indicates the presence of a radiopaque mass 0.6 cm in diameter in the left kidney. Further specialized work-up reveals elevated levels of the amino acids cystine, arginine, lysine, and ornithine in her urine.

1. If the function of the cells of this patient's renal proximal tubules were compared with those of a healthy person, which of the following defects in the biochemistry of cystine, arginine, lysine, and ornithine would likely be exhibited?
   A. Increased synthesis
   B. Excessive secretion
   C. Decreased metabolism
   D. Reduced uptake
   E. Normal uptake, but abnormal re-secretion

2. Defects in glucose uptake into muscle cells are characteristic of insulin resistance in type 2 diabetes and the metabolic syndrome. This phenomenon is likely to be due to reduced activity of a transporter that operates by what mechanism?
   A. Active transport coupled to a sodium-gated channel
   B. Facilitated diffusion followed by phosphorylation
   C. Active transport coupled to ATP hydrolysis

D. Active transport involving antiport with Cl^- and HCO_3^- ions

E. Active transport coupled to outward potassium current

A 4-month-old girl is brought to the pediatrician because of irritability that has led to feeding problems. The parents also are concerned about their daughter's stiff appearance, fits of vomiting, and occasional unexplained fevers. The patient is at the 20th percentile for weight and 25th for height. Physical examination shows weakness and reduced reflexes in the limbs, and there is minimal response to verbal and visual stimulation. A complete blood count is normal. Audiometry suggests bilateral deafness, and an MRI of her head reveals abnormal white matter. Genetic testing indicates a mutation in the gene encoding galactosylcerebrosidase, a lysosomal enzyme.

3. What is the most likely diagnosis for this patient's condition?

A. Pompe disease

B. Gaucher disease

C. Krabbe disease

D. Fabry disease

E. Schindler disease type I

4. Certain drugs are thought to increase membrane fluidity directly, resulting in impaired neurotransmission that may be the basis for their therapeutic effects. Which class of drugs acts by this direct mechanism?

A. Hallucinogens

B. Stimulants

C. Sedatives

D. Opiates

E. Anesthetics

A 27-year-old white man seeks medical attention complaining of "forgetfulness" that has begun to interfere with his ability to work. Lately, he has stumbled over chores at work that he had been doing for years. He has also noticed that the dimensions of his facial features have changed over the past 3-4 years. He brought a 4-year-old photo of himself to show that the bony structures of his chin, cheeks, and forehead have become more prominent and coarser. Physical examination reveals angiokeratomas on his torso. Ultrastructural examination shows that his skin cells have lysosomal inclusions.

5. Biochemical analysis of the lysosomes from this patient's skin cells would likely reveal a deficiency of which of the following enzymes?

A. Glucocerebrosidase

B. Lysozyme

C. α-N-acetylgalactosaminidase (α-NAGA)

D. Galactocerebrosidase

E. α-Galactosidase A

6. Despite the fact that *trans* fatty acids are unsaturated, their contributions to atherosclerosis are similar to those of saturated fats. This similarity in physiologic action can be attributed to which of the following?

   A. Similar rates of metabolism

   B. Relatively linear structures

   C. Similar tissue distributions

   D. Solubilities in water

   E. Tendency to form triglycerides

ANSWERS

1. The answer is D. The patient's symptoms are consistent with a kidney stone, which is confirmed by the radiographic finding. The etiology of the stone is indicated by the urinalysis data, which suggest cystinuria. The cells of this patient's renal proximal tubules would be deficient in a transporter responsible for the reabsorptive uptake of cystine and the basic amino acids, arginine, lysine, and ornithine. Failure of the tubules to reabsorb these amino acids from the ultrafiltrate causes them to be excreted at high concentration in the urine.

2. The answer is B. Glucose uptake by the GLUT4 insulin-responsive glucose transporter in muscle and fat cells operates by passive transport or facilitated diffusion. As such, no energy input derived from ATP hydrolysis or by dissipation of pH or ion gradients is needed for the uptake itself. The glucose concentration gradient is maintained in favor of uptake by rapid, efficient phosphorylation of glucose upon its entry into the cell. Thus, the intracellular glucose concentration at any given time is essentially zero, so there is no need to expend energy for active transport.

3. The answer is C. Of the lysosomal storage disorders listed, Fabry disease can be ruled out because it is X-linked (and thus rarely seen in females) and because of the absence of paresthesias and skin lesions. All the other options would be consistent with the neuromuscular symptoms, ie, weakness and spasticity. However, Gaucher disease is a remote possibility, since no bruising or anemia was noted. Genetic testing provided the key information for the diagnosis; deficiency of galactosylcerebrosidase occurs in Krabbe disease.

4. The answer is E. Anesthetics are highly lipid-soluble and experiments with isolated membranes indicate that these molecules can dissolve in the hydrophobic center of the membrane bilayer. This causes a measurable increase in the membrane fluidity by disrupting the packed structure of phospholipids tails. This is considered to be the main, direct mechanism by which this class of drugs inhibits neurotransmission (pain sensations) in neurons. Hallucinogens and opiates may also affect membrane fluidity, but their effects occur by indirect mechanisms, resulting from changes in the protein or lipid composition of the membranes.

5. The answer is C. The patient's symptoms are consistent with a lysosomal storage disorder of a progressive type. The appearance of features rather late in life encompassing

developmental regression, coarsening facial features, and occurrence of keratomas on the torso are suggestive of Schindler disease type II or Kanzaki disease. This disorder is caused by deficiency of the enzyme α-NAGA, which causes accumulation of glycosphingolipids in the lysosomes, corresponding with the inclusion bodies observed in microscopic examination of the patient's cells. The other enzymes listed are involved in various storage diseases, but their characteristics are readily distinguished from Kanzaki disease.

6. The answer is B. Saturated fatty acids and *trans* fatty acids are structurally similar; their hydrocarbon tails are relatively linear. This allows them to pack tightly together in semi-crystalline arrays such as the membrane bilayer. Such arrays have similar biochemical properties in terms of melting temperature (fluidity). Although some of the other properties listed are also shared by saturated and *trans* fats, they are not thought to account for the tendency of these fats to contribute to atherosclerosis.

CHAPTER 5
METABOLIC INTERRELATIONSHIPS AND REGULATION

I. Diet and Nutritional Needs

A. Nutrients taken into the body via the diet can have different metabolic fates—**catabolism** or **anabolism.**

1. **Catabolism** refers to metabolic processes by which nutrient molecules are **degraded to simple products** (waste) in order to extract **energy.**

   a. Catabolic processes operate in stages.

   (1) The first step is to hydrolyze polymeric nutrient molecules to their component **building blocks,** eg, polysaccharides to simple sugars.

   (2) The second step involves "burning" or **oxidation** of their carbon skeletons to extract electrons, from which energy can be derived through formation of ATP.

   b. Catabolism predominates when the body's energy stores are low and need to be replenished.

2. **Anabolism** encompasses the **synthesis** of complex macromolecules and structures from building blocks derived from nutrients as well as synthesis of the building blocks themselves, such as nonessential amino acids.

   a. These macromolecules include cellular proteins and nucleic acids as well as storage forms of fuels, eg, glycogen and triacylglycerols.

   b. These synthetic processes are critical for **maintenance of organ function** by replacing proteins that have been degraded and by enabling cell division and differentiation.

   c. These are major **energy-requiring** processes, which can proceed only when energy and fuel reserves are abundant.

3. Catabolism and anabolism are often **inversely regulated** to provide balance for maintenance of the body's basal metabolic rate and to enable specific physiologic functions of organs.

B. Nutritional balance and dietary intake have a major impact on the health of human populations.

1. **Overnutrition** in developed nations has led to major health problems with epidemic **type 2 diabetes mellitus** and **obesity.**

2. **Undernutrition** arising from poor quality or limited availability of food in developing nations has produced conditions of starvation and malnutrition.

PROTEIN-CALORIE MALNUTRITION

- *Most cases of protein-calorie malnutrition in the United States are secondary to a highly catabolic condition, such as trauma or a major infection.*
- *In countries where food is in short supply or the diet is inadequate, protein-calorie malnutrition can take two extreme forms, kwashiorkor and marasmus.*
- ***Kwashiorkor*** *arises in children due to **deprivation of protein relative to calories,** eg, a starch-dominated diet.*
  - *Symptoms and effects include stunted growth, edema, dermal lesions, loss of hair pigmentation, and decreased plasma albumin.*
  - *Fat deposition leads to visible enlargement of the liver, resulting in **distended abdomens** that are characteristic in afflicted children.*
- ***Marasmus*** *occurs as a result of **deprivation of calories relative to protein,** eg, a diet mainly of milk.*
  - *Symptoms include arrested growth, extreme **muscle wasting (emaciation),** weakness, and anemia; all these symptoms contribute to frequent infections.*
  - *The absence of edema or reduction in albumin distinguishes marasmus from kwashiorkor.*

C. The main role of **dietary proteins** is provision of the amino acid building blocks for synthesis of cellular proteins, many of which require daily renewal to maintain physiologic functions and respond to the needs of the body.
1. **Digestive enzymes** must be produced in large quantities each day.
2. **Protein turnover** is a natural process resulting from the balance between degradation and synthesis.
3. Protein synthesis is required for production of new cells to replace those lost to normal turnover, such as skin cells and RBCs.
4. **Nitrogen balance** is determined by how well the amount of dietary nitrogen-based compounds (principally proteins) matches the nitrogen needs of the body.
    a. In **positive nitrogen balance,** more nitrogen-based nutrients are taken in than needed.
       (1) In this metabolic condition, adequate nitrogen-containing compounds are available for the reactions that require them.
       (2) Any excess protein intake is converted for use of the **carbon skeletons of the amino acids as energy** and the amino groups are excreted as urea (see Chapter 9).
       (3) If total caloric intake exceeds the energy needs of the body, then the carbon skeletons may be converted for **storage as fat** (see Chapter 8).
    b. **Negative nitrogen balance** occurs when more nitrogen is excreted than taken in.
       (1) This is characteristic of **starvation** and **disease states,** such as chronic infection or cancer.
       (2) These conditions may produce **cachexia,** in which increased degradation of proteins leads to **muscle wasting.**
5. Amino acids that cannot be synthesized by the body are termed **essential amino acids,** and a diet deficient in even one essential amino acid can lead to negative nitrogen balance.
D. Most dietary **carbohydrates** are digestible, ie, capable of being metabolized and used for energy by the body.
1. Digestible carbohydrates include **simple sugars,** disaccharides, and polysaccharides (such as **starches**).

2. Carbohydrates are mainly used as fuel, either in a direct manner, after storage as glycogen, or after conversion to lipids.
   a. The main pathway for **glucose metabolism** in the presence of oxygen involves dismantling the sugars via **glycolysis,** extracting electrons from them by the enzymes of the tricarboxylic acid (TCA) cycle, and using those electrons to **produce ATP by the electron transport chain** (see Chapters 6 and 7).
   b. Some dietary sugars are used to replenish supplies of **glycogen,** a **polymer of glucose** that is the main storage form of the sugar in the body, primarily in the liver and skeletal muscle (see Chapter 6).
   c. Once the energy needs of the body are met and glycogen stores have been replenished, remaining sugars are **converted to fat,** ie, **triacylglycerol** for storage, mainly in adipose tissue (see Chapter 8).
3. Dietary sugars are also modified for synthesis of **glycoproteins** and **proteoglycans,** especially for serum proteins and extracellular matrix structural proteins.

E. Lipids or fats have structural or signaling functions in the body, in addition to their major role in energy storage.
1. Dietary fats have a **very high energy content.**
   a. Complete burning of fats to CO_2 and H_2O via aerobic metabolism produces 9 kilocalories per gram, compared with 4 kilocalories per gram from carbohydrates or proteins.
   b. This property makes fats the most efficient storage form of energy reserves in the body.
2. Fats are very important for function of cell membranes.
3. Fats are also used for synthesis of specialized signaling molecules, such as **prostaglandins, thromboxanes,** and **leukotrienes.**

II. Regulation of Metabolic Pathways

A. Many metabolic pathways are regulated by **allosteric control** of **key enzymes catalyzing the rate-limiting step** of the pathway (Figure 5–1).
1. **Anabolic pathways** are frequently stimulated under conditions of abundance (ie, high levels of cellular energy and availability of precursor molecules for the pathway) and inhibited when energy and precursors are low.
2. **Catabolic pathways** are often activated by conditions involving low energy and are inhibited when energy and building blocks are available at high levels.

B. The **rate** or **flux** of substrates through a pathway is also dependent on **substrate availability.**

C. One of the major mechanisms for regulation of preexisting enzymes is via **covalent modification,** usually by protein **phosphorylation** or **dephosphorylation.**
1. This is an important mechanism because it is rapid, reversible, and economical for the cell and body.
   a. These changes can be implemented within seconds or minutes, allowing a quick response to environmental stimuli.
   b. This mechanism saves energy because such changes **do not require new protein synthesis** or **altered gene expression** to affect activity of a protein or enzyme.
   c. Reversal of the phosphorylation state restores the original condition without the cost of degrading and replacing the protein.

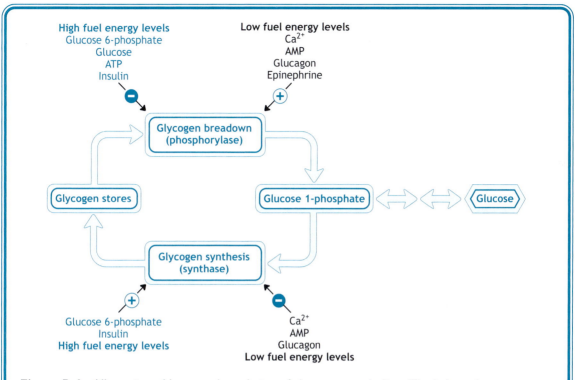

Figure 5–1. Allosteric and hormonal regulation of glycogen metabolism. The balance between synthesis (anabolism) and breakdown (degradation or catabolism) is regulated by molecules that reflect the energy status of the cell. Allosteric control is often reciprocal, eg, glucose 6-phosphate. In most cases, the actions of insulin are reciprocal to those of glucagon and epinephrine, eg, insulin action activates glycogen synthase while inhibiting the activity of phosphorylase in the same cells. Hormones initiate their actions by binding to their cognate receptors. This leads to activation of kinases or phosphatases that modify the activities of glycogen synthase and phosphorylase by phosphorylation-dephosphorylation events.

    2. Addition or removal of a phosphate group alters intrinsic protein activity through changes in conformation.

      a. These effects can either be **stimulatory** or **inhibitory.**

      b. **Kinases** (phosphotransferases) add phosphate; **phosphatases** (phosphohydrolases) remove it.

D. **Long-term metabolic responses,** occurring over the course of hours or days, involve **regulation of gene expression** to induce changes in levels of one or more enzymes.

    1. These effects are produced by altered **transcription** of genes through changes in activity of signaling pathways leading to transcription factors, which leads to corresponding changes in **protein synthesis** (see Chapter 12).

    2. These effects can also arise from differences in **rates of degradation** or turnover of the finished proteins.

E. **Hormonal control** provides a major means for regulation of metabolic pathways, involving the opposing actions of insulin *versus* glucagon or epinephrine (Figure 5–1).
   1. **Insulin** is the **anabolic hormone** secreted by the **beta cells of the pancreatic islets of Langerhans** in response to increases in blood levels of glucose, amino acids, and fats after a meal.
      a. Insulin action **promotes storage** of sugars, amino acids, and fats and **stimulates synthesis** of macromolecules (eg, proteins) from simple precursors.
      b. Conversely, insulin action inhibits the pathways involved in breakdown of macromolecules.
      c. These actions of insulin are mediated by **reversible phosphorylation/dephosphorylation** events in the short-term and can also alter **gene expression** over a period of hours (see Chapters 6 and 14 for further details).
   2. **Glucagon,** secreted by the **alpha cells of the islets of Langerhans,** is the main **catabolic hormone.**
      a. Glucagon action promotes usage of glucose and alternative fuels by many tissues and **stimulates net degradation** of macromolecules to provide energy and to **increase blood glucose** levels.
      b. Glucagon also **inhibits** many of the **synthetic pathways** in order to spare energy for critical cellular and bodily functions.
      c. The actions of glucagon are mediated by reversible **phosphorylation/dephosphorylation** through the actions of cyclic AMP–dependent protein kinase (see Chapter 14), which can alter enzyme activities in a rapid manner and affect gene expression in the long-term.
   3. **Catecholamines,** such as **epinephrine** secreted by the chromaffin cells of the adrenal medulla or **norepinephrine** produced by the pancreas, have similar actions on metabolism to those of glucagon.
      a. Epinephrine release usually occurs in response to stress—the rapid "**fright, fight or flight**" response.
      b. Like glucagon, the actions of catecholamines are partially mediated through increased **cyclic AMP** levels and altered protein phosphorylation.
      c. Part of the similarity in action is also due to increased glucagon secretion induced by the catecholamines.

III. Glucose Homeostasis

   A. Maintenance of blood glucose within a narrow concentration range is critical to proper bodily function.
   1. **Glucose is required as the sole fuel** for certain tissues, especially the **brain.**
   2. The **liver** and **kidney** are the main organs involved in regulating blood glucose, both directly in response to blood glucose rise and fall as well as in response to hormones.

   B. Regulation of blood glucose concentration occurs initially through changes in its uptake and phosphorylation to **glucose 6-phosphate**.
   1. Glucose uptake is mediated by the **glucose transporters GLUT1** or **GLUT4** followed by phosphorylation by hexokinase.
   2. Glucose uptake and phosphorylation mechanisms in the liver respond to meet the body's needs as blood glucose concentrations rise and fall over the course of the day.
      a. The **liver glucose transporter, GLUT2,** can operate in **two directions.**

(1) Glucose is taken up into liver cells when available in abundance in blood, eg, after a meal.

(2) Glucose derived from glycogen stores or made in gluconeogenesis is transported out of liver cells to increase blood glucose when the external concentration is low.

b. The enzymes responsible for phosphorylation of glucose to glucose 6-phosphate, **hexokinase** and **glucokinase,** have distinct kinetic properties that allow the liver to respond to increased glucose availability after a meal (see Chapter 6 for details).

C. Blood glucose is regulated by the hormones insulin, glucagon, and epinephrine.

1. The **actions of insulin** and **glucagon** (or epinephrine) on the complex processes that contribute to regulation of blood glucose **oppose each other** (Figure 5–2).

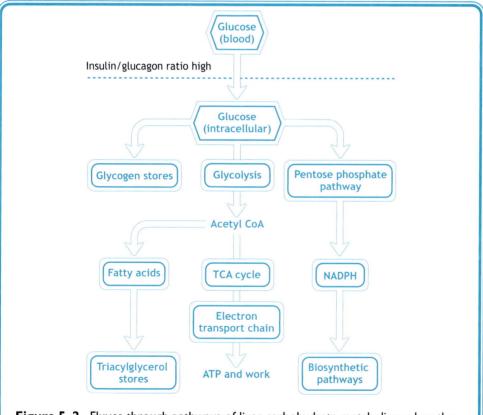

Figure 5–2. Fluxes through pathways of liver carbohydrate metabolism when the insulin/glucagon ratio is high. Blood glucose is elevated after a meal, and some of this fuel is stored as glycogen for later use. The remainder either may be metabolized for immediate generation of energy (ATP) or to produce reducing equivalents (NADPH) needed for synthesis of fatty acids, nucleic acid building blocks, and other compounds. Acetyl CoA produced from glucose in excess of energy needs can be converted to fatty acids for storage in adipose tissue as triacylglycerols.

2. **Glucagon** and **insulin secretion** is regulated in response to **blood glucose** levels.
   a. The pancreatic glucose transporter **GLUT2** is the **glucose sensor**.
   b. When glucose is high, insulin secretion is stimulated and glucagon secretion is inhibited.
   c. When glucose is low, insulin secretion is inhibited and glucagon secretion is stimulated.
3. **Insulin** action on carbohydrates is mainly designed to **decrease blood glucose** (Figures 5–2 and 5–3).
   a. In the liver and kidney, insulin stimulates glucose uptake as well as glycolysis and glycogen synthesis and simultaneously suppresses glycogen degradation and gluconeogenesis.
   b. In muscle, insulin stimulates glucose uptake and utilization, as well as glycogen synthesis, and inhibits glycogen degradation.
   c. In adipose tissue, insulin stimulates glucose uptake and utilization.
4. **Glucagon** action on carbohydrates is designed to **increase blood glucose** levels (Figure 5–3).
   a. In the liver, glucagon stimulates glucose production by glycogenolysis and gluconeogenesis.
   b. There is no effect of glucagon on glycogen metabolism in muscle.

IV. Metabolism in the Fed State

A. **Digestive enzymes** of the **gastrointestinal tract** begin **hydrolysis** of protein, fat, and carbohydrates into their component building blocks, namely, amino acids, fatty acids and monoacylglycerols, and simple sugars (such as glucose).
   1. Intestinal **epithelial cells take up** these compounds, process them further, and then release them into the hepatic portal circulation.
   2. **Increased blood levels** of these nutrients, especially **glucose** and **amino acids,** stimulate the pancreas to release insulin and suppress glucagon release.
   3. During the **absorptive** or **fed state** (up to 2–4 hours after a meal), metabolic events in the body allow processing of the food-derived compounds (Figure 5–4).
      a. **All tissues utilize glucose for energy** during this time.
      b. The **high insulin/glucagon ratio** stimulates **anabolic processes** in many organs.
B. The **liver** is the first organ to respond to the influx of nutrients after a meal.
   1. The hepatic portal vein carries the nutrients directly to the liver.
   2. The liver takes up these nutrients and then metabolizes them or targets them to be stored.
   3. Glucose is taken up and phosphorylated mainly by glucokinase, which initiates several processes of glucose utilization.
      a. The rate of **glycolysis increases,** which allows glucose metabolism to provide energy for the organ.
      b. **Net storage of glucose as glycogen** is due to stimulation of glycogen synthesis and inhibition of its breakdown.
      c. Some glucose is metabolized by the **pentose phosphate pathway** to produce NADPH for use in biosynthetic reactions by the liver.
      d. **Gluconeogenesis,** the pathway for synthesis of new glucose and one of the major functions of the liver, is **inhibited** during this time.
   4. **Synthesis of fatty acids** and their incorporation into triacylglycerols are **stimulated** during this time of energy excess.

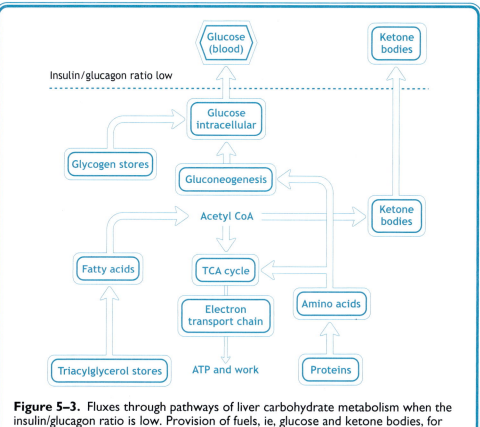

Figure 5–3. Fluxes through pathways of liver carbohydrate metabolism when the insulin/glucagon ratio is low. Provision of fuels, ie, glucose and ketone bodies, for use by other tissues is the main goal under these conditions. Glycogen stores provide glucose during the first 24 hours of fasting. The carbon skeletons of amino acids from protein degradation and glycerol and fatty acids from breakdown of triacylglycerols provide the raw materials for fuel production in long-term fasting. TCA, tricarboxylic acid.

    5. Amino acid levels are also **elevated in the blood** after a meal, and this wealth of raw materials is **managed by the liver** in one of several ways.
       a. Amino acids are used directly by the liver for **synthesis of new proteins.**
       b. Some of the excess amino acids are released into the bloodstream for **utilization by other tissues.**
       c. Alternatively, the excess amino acids are **metabolized** to store their **carbon skeletons** for later use to produce energy.
  C. Adipose is the tissue where much of the body's **energy reserves are stored as fats,** specifically triacylglycerols, so its role after a meal is to **convert any excess fuel to fat.**
    1. Insulin action increases glucose uptake by individual fat cells (adipocytes), and this accelerates metabolic activity.
       a. The rate of **glycolysis is increased** to provide **energy,** acetyl CoA, and glycerol 3-phosphate to be used to make triacylglycerols.

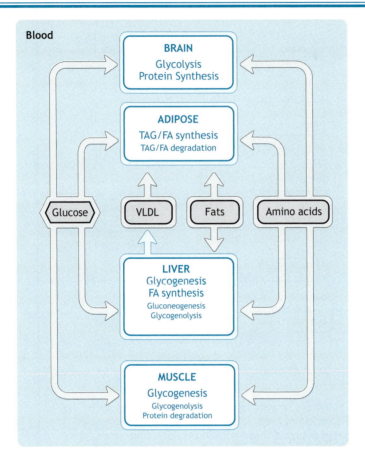

Figure 5–4. Metabolic activities of major organs in the fed state. The relative activities of major metabolic pathways or processes in each of the organs are indicated by their font sizes. The exchange of nutrient materials and fuel molecules through the bloodstream illustrates the interrelationships of these organs. In the absorptive condition, all organs share the bounty of nutrients made available by digestion of food by the intestine. PPP, pentose phosphate pathway; FA, fatty acids; TAG, triacylglycerol.

    b. The **pentose phosphate pathway is stimulated to produce NADPH,** which may be needed later for fatty acid synthesis.

  2. There is net **synthesis of triacylglycerols** for storage.

    a. **Free fatty acids** delivered by the bloodstream and derived from dietary fats are attached to a glycerol backbone for **storage as triacylglycerol** in the large fat droplet of each adipocyte.

    b. Breakdown of the stored triacylglycerols is inhibited at this time.

D. **Skeletal muscle utilizes and stores glucose** in the fed state.

  1. As it does in adipose tissue, insulin promotes **increased glucose uptake** by skeletal muscle.

a. The glucose is converted to glucose 6-phosphate by hexokinase and some is metabolized through **glycolysis** and **oxidative phosphorylation for energy.**

      b. The **glycogen stores of muscle are not extensive** and can be depleted within a few minutes of intensive exercise, but the high level of glucose 6-phosphate availability after a meal allows **glycogen synthesis to replenish the stores.**

    2. Insulin action and the availability of adequate energy and amino acids **stimulate net synthesis of muscle protein,** with suppression of protein degradation.

  E. The **fuel needs of the brain** are both large and of very high priority.

    1. Glucose is the sole fuel for the brain, and this need is easily met in the absorptive state.

    2. There are no stores of glycogen or triacylglycerols in the brain.

OBESITY—DYSREGULATION OF FAT METABOLISM

CLINICAL CORRELATION

- *Nearly two-thirds of Americans are classified as overweight according to the criteria of **body mass index (BMI)** calculations, and obesity is now considered to be a disease.*
  - *In simple terms, weight gain occurs when **calorie intake exceeds calorie usage,** and the excess fuel is stored as fat.*
  - *A **sedentary lifestyle** and the availability of abundant amounts of energy-dense foods are important contributing factors to epidemic obesity in the United States and in many areas of the developed world.*
- *Major sequelae of obesity include **increased risk of type 2 diabetes, hypertension, heart disease** (collectively, the metabolic syndrome or syndrome X), certain cancers, fatty liver and gallstones, arthritis and gout, with attendant reduction in life expectancy.*
- ***Abdominal or visceral fat cells** have a higher rate of fat turnover and are more contributory to disease than fat stores in the buttocks and thighs.*
  - *Fatty acids released from visceral fat move through the hepatic portal circulation directly to the liver, leading to altered hepatic fat metabolism.*
  - ***Dyslipidemia,** characterized by low blood levels of HDL and elevated LDL, leads to atherosclerosis and heart disease.*
  - *Obesity in children has even more devastating long-term consequences because their adipocytes respond to the excess storage demands by dividing to produce more visceral adipocytes, which increases the lifetime storage capacity.*
- ***Adipose is an endocrine gland** that secretes a variety of factors that have effects both in the brain and the peripheral insulin-responsive tissues.*
  - *Adipocytes secrete leptin, adiponectin, and resistin, whose mechanisms of action to mediate peripheral insulin resistance are not yet fully understood.*
  - *Investigations to understand the metabolic changes caused by obesity are in progress, but it is clear that many of the consequences are due to altered signals arising from the increased mass of adipose tissue.*
- *The main treatment for obesity involves lifestyle alteration (ie, decreased caloric intake coupled with increased exercise); however, in severely obese patients, **gastric bypass surgery** is a viable alternative.*

V. Metabolism in the Fasting State

  A. During the **post-absorptive** or **fasting state** (4–24 hours after the last meal), **blood glucose levels begin to fall,** precipitating major changes in metabolism with a switchover from an anabolic state to a catabolic condition in order to maintain blood glucose levels (Figure 5–5).

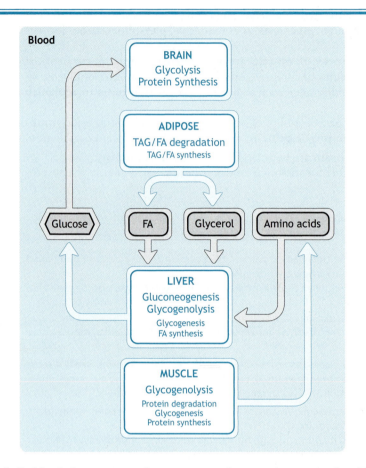

Figure 5–5. Metabolic activities of major organs during a short-term fast. The importance of the liver in providing glucose to support the brain and other glucose-requiring organs in the post-absorptive state is illustrated. The body relies on available glycogen stores as a ready source for glucose as fuel. PPP, pentose phosphate pathway; FA, fatty acids; TAG, triacylglycerol.

1. Insulin levels in the blood decline.
2. Glucagon levels increase.
3. The **decreased insulin/glucagon ratio** activates degradation of glycogen, protein, and triacylglycerols.
4. Most biosynthetic pathways slow down.
5. Gluconeogenesis is stimulated.

B. In its critical role as the central organ for synthesis and distribution of fuel molecules, the **liver** is mainly focused on **export of glucose to peripheral tissues** during a short-term fast.

1. The decreased insulin/glucagon ratio leads to **inhibition of glycogen synthesis** and **increased glycogenolysis** to supply some of the body's glucose needs on an immediate basis.

2. Glycolysis decreases and gluconeogenesis increases.

3. The combination of these effects leads to increased intracellular glucose concentration, much of which is exported from the liver via reversal of transport mediated by GLUT2.

4. During the fasting state, the energy needs of the liver are provided by **fatty acid catabolism** (β-oxidation), which spares further glucose for export to peripheral tissues.

C. **In adipose tissue, reduced glucose availability** via the blood and the low insulin/glucagon ratio lead to **net degradation of triacylglycerols** to their component fatty acids and glycerol to meet the energy needs of most tissues (with the notable exception of the CNS).

1. The fatty acids are oxidized to provide for the energy needs of the adipocytes themselves.

2. As the fast progresses, more of the adipose-derived fatty acids are transported in the bloodstream as **complexes with albumin** and taken up by the **liver.**

3. The glycerol backbones from triacylglycerol breakdown are sent to the liver for use in gluconeogenesis.

D. **Skeletal muscle** in its resting state can satisfy most of its energy needs by **oxidation of fatty acids** taken up from blood, and during the early stages of fasting, **protein degradation in the muscle is increased.**

1. Up to one-third of muscle protein may be degraded to component amino acids for use as fuel during fasting.

2. Most of these **amino acids are released** into the bloodstream and **taken up by the liver** and used as a major **source of fuels.**

   a. Some of the carbons skeletons derived by removal of the amino groups from the amino acids can be used for synthesis of glucose via **gluconeogenesis.**

   b. Some carbon skeletons yield acetyl CoA and are used for synthesis of the alternative fuel, **ketone bodies,** which become more important as the fast extends past 24 hours.

3. Glycogen stores in skeletal muscle are mainly held in reserve to satisfy the organ's need for a burst of energy during exercise, and thus are rapidly depleted upon activity during a fast.

E. The **energy needs of the brain** and other glucose-requiring organs are satisfied during the post-absorptive period through **provision of glucose by the liver.**

VI. Metabolism During Starvation

A. If **fasting extends past 1–2 days,** which is considered to be a long-term fast or **starvation,** further changes in fuel synthesis and use by several organs can occur, principally a **conversion from a glucose economy to** one dominated by **ketone bodies as fuel** (Figure 5–6).

1. In addition to the effects of a low insulin/glucagon ratio, long-term changes in metabolism during starvation are induced by the corticosteroid, **cortisol.**

2. Cortisol promotes **net protein breakdown** in skeletal muscle to provide amino acids as precursors for gluconeogenesis and ketone body synthesis (**ketogenesis**).

3. Cortisol also increases the rate of triglyceride breakdown (**lipolysis**) in adipose tissue for these same purposes.

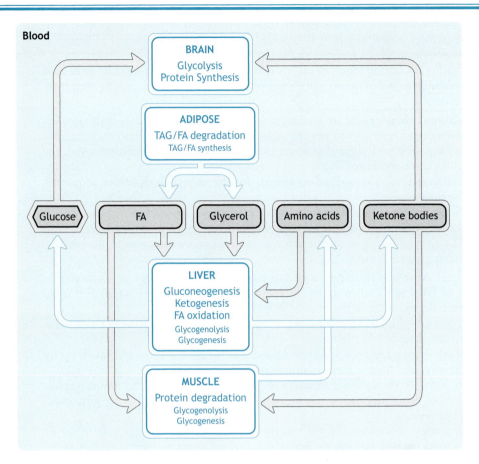

Figure 5–6. Metabolic activities of major organs during long-term fasting. With glycogen stores in the liver and muscle depleted, gluconeogenesis is the sole means of providing for the glucose needs of some organs, while many organs, even the brain, adapt to use of the alternative fuel, ketone bodies, which is derived mainly from degradation of fatty acids. FA, fatty acids; PPP, pentose phosphate pathway; TAG, triacylglycerol.

B. The **liver** is again the major organ that synthesizes the principal long-term fuel, **ketone bodies,** acetoacetate, and 3-hydroxybutyrate, which are made from both amino acids and fatty acids.

C. In prolonged fasting, **triacylglycerol degradation in adipose tissue becomes maximal** and sustained.

D. Protein breakdown in skeletal muscle can only be sustained for 10–14 days, at which point further degradation of protein would severely compromise contractile capability.

E. Within a few days of fasting, the **brain adapts to be able to utilize ketone bodies as fuel** and becomes less dependent on, but never completely independent of, glucose.

TYPE 1 DIABETES MELLITUS

- *Patients with type 1 diabetes (previously called juvenile or insulin-dependent diabetes) have an **absolute deficiency of insulin,** which produces **chronic hyperglycemia** (elevated blood glucose) with elevated **risk for ketoacidosis** and a variety of long-term complications, including retinopathy, neuropathy, nephropathy, and cardiovascular complications.*
  - *Even in persons with well-controlled diabetes, the **long-term complications** include stroke, heart attack, renal disease, blindness, and limb amputation.*
  - *Onset of type 1 diabetes mellitus usually occurs within the first two decades of life; presenting symptoms include hyperglycemia, polyuria, polydipsia, and polyphagia (excessive urination, thirst, and appetite, respectively), often with serious ketoacidosis in response to a stressor such as a viral infection.*
  - *The diagnosis may be supported by an abnormal glucose tolerance test.*
- *The etiology of type 1 diabetes is **autoimmune destruction of the pancreatic beta cells,** which is initiated by an event such as viral infection and progresses to the point of frank symptoms during childhood and the teenage years.*
  - *Evidence suggests a genetic predisposition toward the autoimmune response, but the genes involved are unknown.*
  - *At this time, it is not possible to diagnose the disease prior to appearance of symptoms, nor is there a way to stop its progression.*
- *The **metabolic disruption** in type 1 diabetes is due to both the **absence of insulin action and unopposed glucagon action** in liver, muscle, and adipose tissues.*
  - *Failure of insulin to suppress gluconeogenesis in liver leads to overproduction of new glucose, which exacerbates the elevation of blood glucose due to decreased uptake of dietary glucose by muscle and adipose.*
  - *In the absence of insulin and in response to glucagon stimulation, triacylglycerol degradation in adipose tissue runs unabated and the flood of fatty acids reaching the liver leads to ketone body synthesis and packaging of some triacylglycerols into VLDLs.*
  - *In some ways, the metabolic profile of a patient with uncontrolled type 1 diabetes resembles that of the starved patient, except that in the complete absence of insulin, the ketoacidosis of diabetes is much more severe than in fasting, and starvation is rarely associated with hyperglycemia.*
- *Peripheral tissues (such as liver, skeletal muscle, and adipose) retain normal responsiveness to insulin, and management of the disease involves subcutaneous **insulin injection** with monitoring of blood glucose several times per day.*
  - *Standard treatment involves one or two daily injections of a prescribed dose of insulin, which is less likely to produce hyperinsulinemia leading to episodes of hypoglycemia.*
  - *At best, standard treatment brings blood glucose levels down to about 140–150 mg/dL (normal = 110 mg/dL).*
  - *However, elevated glucose over many years inevitably produces the debilitating complications of the disease through **protein glycation** events (ie, addition of glucose to proteins, especially those lining blood vessels, leading to protein dysfunction).*
  - *Intensive treatment involves a more aggressive attempt to manage blood glucose levels by monitoring blood glucose multiple times during the day and administration of six to eight small doses of insulin as needed.*
  - *Another method for aggressive control of blood glucose levels is the use of **insulin pumps** to cover basal insulin needs plus supplemental dosing at meals with fast-acting insulin.*
  - *The benefit of this approach is decreased blood glucose to reduce the risk of long-term complications, but the main drawback of intensive treatment is possible overdosing producing hypoglycemia, which may cause disorientation, loss of consciousness, coma, and death.*
  - *Hypoglycemic agents, which are an important part of the therapeutic repertoire for type 2 diabetes, do not work in cases of type 1 diabetes.*
- *There are approximately 1 million cases of diagnosed type 1 diabetes mellitus in the United States.*

TYPE 2 DIABETES MELLITUS

- Type 2 diabetes is by far the **more prevalent form** of diabetes in the United States, with ~10 million diagnosed cases, and new cases are being diagnosed at an increasing rate of > 600,000 per year.
- The disease is characterized by **peripheral insulin resistance** leading initially to increased secretion of insulin by the pancreatic beta cells.
  - Chronic overwork eventually leads to beta cell dysfunction, and insulin secretion becomes inadequate to maintain blood glucose with development of symptoms.
  - Although the exact molecular basis for the insulin resistance is not known, there are strong associations with obesity and a sedentary lifestyle.
  - There is a **very strong genetic component** to type 2 diabetes, with evidence favoring a polygenic disease mechanism but with few of these genes definitively identified.
- The symptoms of type 2 diabetes include **hyperglycemia without the ketosis** associated with type 1 disease due to residual effects of insulin on ketone body synthesis.
  - **Hypertriacylglycerolemia** with secretion of increased VLDL can lead to long-term elevated risk of atherosclerosis, although this is a complicated, multifactorial process.
  - Other long-term complications are similar to those caused by type 1 diabetes, likely due to the chronic hyperglycemia.
- Treatment of type 2 diabetes, at least in its early stages, mainly involves lifestyle modification.
  - Recommendations include a **calorie-restricted diet** and **increased exercise,** with the goal of weight reduction.
  - Significant weight reduction can actually resolve the insulin resistance in some patients.
  - Insulin injections are not normally needed to manage blood glucose levels in persons with type 2 diabetes, except in those with advanced-stage disease when pancreatic insulin production is extremely low and patients benefit from supplemental insulin.
- When lifestyle changes alone are insufficient to manage blood glucose levels, a variety of **hypoglycemic agents** can be used.
  - **Sulfonylureas,** such as glipizide and glyburide, and **meglitinides,** such as repaglinide and nateglinide, stimulate insulin secretion by the beta cells.
  - **Biguanides,** such as metformin, suppress liver gluconeogenesis and enhance insulin action in muscle.
  - **Thiazolidinediones,** such as pioglitazone and rosiglitazone, reduce blood glucose levels by enhancing glucose utilization in response to insulin in adipose and muscle and decreasing gluconeogenesis in the liver.
  - **α-Glucosidase inhibitors,** such as acarbose and miglitol, block hydrolysis of dietary starches and thereby reduce dietary glucose absorption.

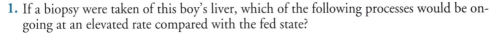

CLINICAL PROBLEMS

A 15-year-old boy awakens at 7:30 AM and as he sits down at the breakfast table, he exclaims that he "is really starving." The boy finished dinner at 7:15 PM the previous evening and had not remembered to have a snack before going to bed.

1. If a biopsy were taken of this boy's liver, which of the following processes would be ongoing at an elevated rate compared with the fed state?

   A. Protein synthesis

   B. Glycogenolysis

C. Glycolysis

D. Fatty acid synthesis

E. Pentose phosphate pathway

2. The insulin resistance that is the hallmark of type 2 diabetes mellitus is thought to arise from multiple factors. Of the putative contributing factors listed below, which is likely to be the most direct contributor to the disease?

A. Endocrine signals from the visceral adipose

B. Death of pancreatic beta cells

C. Increased mass of adipose in thighs and buttocks

D. Dysfunction of lipid metabolism in liver

E. Sedentary lifestyle

A student finished eating a well-balanced, 750-kilocalorie meal just 1 hour ago and has since been sitting quietly watching television.

3. Which of the following substances would NOT be elevated in this student's blood?

A. Fatty acids

B. Insulin

C. Amino acids

D. Glucagon

E. Glucose

A 22-year-old woman engaging in a political protest goes on a hunger strike on a prominent corner in a city park. Although food is offered to her several times each day by social workers and the police, she refuses all offers except for water through the first 2 weeks.

4. An examination of a sample of this woman's brain tissue would reveal that her brain had adapted to using which of the following as fuel?

A. Glycerol

B. Amino acids

C. Glucose

D. Ketone bodies

E. Free fatty acids

A 14-year-old girl is brought to the clinic by her father with a complaint of light-headedness experienced on the soccer field earlier in the afternoon. She stated that she felt cold and nearly fainted several times, and that the symptoms did not resolve even after she drank a power beverage. On further questioning, her father stated that she had been very thirsty recently, which bothered him because it meant having to make frequent bathroom stops while driving on trips. She also "eats like a horse" and never seems to gain any weight or grow taller. Physical examination reveals a thin girl who is at the 30th percentile for height and weight. A rapid dipstick test reveals glucose in her urine.

5. Evaluation of this girl's liver would reveal an increased rate of which of the following processes?

   A. Glycolysis

   B. Glycogenesis

   C. Ketogenesis

   D. Fatty acid synthesis

   E. Protein synthesis

ANSWERS

1. The answer is B. After an overnight fast (~12 hours), the liver would be active in secreting glucose derived mainly from breakdown of stored glycogen, but also via gluconeogenesis from amino acid carbon-skeleton precursors. All the other processes listed would be decreased relative to the fed state, in order to focus energy on meeting the glucose needs of dependent organs such as the brain. This is especially true of the anabolic processes like the pentose phosphate pathway and pathways for synthesis of fatty acids and proteins. Actually, the liver would be meeting its own energy needs mainly through fatty acid oxidation at this time, which would reduce flux through glycolysis.

2. The answer is A. Recent research has revealed that excess visceral fat deposits secrete several factors that have direct effects on the brain as well as directly on muscle to produce peripheral insulin resistance. Some of these newly identified factors are leptin, resistin, and adiponectin, whose mechanisms of action are still under active investigation. Death of pancreatic beta cells is a hallmark feature of type 1 diabetes and may occur only in very advanced stages of type 2 diabetes. Excess adipose in the thighs and buttocks does not contribute as strongly to insulin resistance as does visceral fat, presumably due to a lower level of endocrine activity of such fat depots. Dysfunction of liver lipid metabolism is more a consequence of excess activity of adipose than a cause of insulin resistance. A sedentary lifestyle contributes to build-up of excess fat stores but does not act directly to induce insulin resistance.

3. The answer is D. This student is still in the fed or absorptive state within 1 hour of a meal, so elevated levels of many nutrients derived from food digestion would be observed in her blood. This would include all items in the list except glucagon. High nutrient levels in the blood evoke increased insulin secretion from the beta cells and suppression of glucagon secretion by the alpha cells of the islets of Langerhans. Therefore, blood levels of glucagon would be decreased relative to other nutritional states.

4. The answer is D. This woman has created a self-imposed starvation through her hunger strike. During starvation, many fuel sources are recruited to support bodily functions, including protein degradation, which supplies amino acids as gluconeogenic precursors, and triacylglycerol degradation, which yields glycerol, free fatty acids and, eventually, ketone bodies. The brain normally prefers glucose as its main fuel, so no adaptation is needed. During starvation, changes in brain gene expression up-regulate

several enzymes to enable use of ketone bodies as fuel. No matter how long the fast lasts, the brain cannot use glycerol, amino acids, or free fatty acids as direct fuel sources.

5. The answer is C. This girl's symptoms are consistent with extreme hyperglycemia, which is consistent with her excessive thirst (polydipsia), urination habits (polyuria), and appetite (polyphagia). Her neurologic symptoms are probably secondary to ketoacidosis, likely resulting from type 1 diabetes. The finding of glucose spillover into her urine strongly supports this conclusion. An acute hyperglycemic condition due to type 1 diabetes is characterized by a near-absence of insulin with unopposed glucagon action, particularly in the liver. So both gluconeogenesis and ketogenesis are elevated in such patients. All the other processes listed would be operating at reduced activity relative to their levels in the presence of a higher insulin-glucagon ratio.

CHAPTER 6
CARBOHYDRATE METABOLISM

I. Digestion and Absorption of Dietary Carbohydrates

A. The main sites of breakdown of dietary carbohydrates are the mouth and the duodenum.

1. The process starts in the mouth during **mastication** where **salivary α-amylase** cleaves some of the α-1,4 glycosidic bonds of **starch.**

2. This process is completed in the **duodenum** where **pancreatic α-amylase** produces a mixture of monosaccharides, disaccharides, and oligosaccharides.

3. Disaccharides are cleaved to monosaccharides by a battery of **disaccharidases** after absorption into intestinal mucosal cells.

  a. For example, sucrose is hydrolyzed to glucose and fructose by **sucrase.**

  b. **Lactase,** which is responsible for hydrolyzing lactose to glucose and galactose, is expressed at low levels in many adults, especially those with **lactose intolerance.**

B. **Uptake** of monosaccharides and disaccharides by intestinal mucosal cells is mediated by a variety of **transporters.**

II. Glycolysis

A. **Glycolysis** is the process by which **glucose** is broken down to **pyruvate** in order to begin obtaining some of the **energy** stored in the glucose molecule for use by the body.

1. The energy released in this process results in the direct formation of **ATP.**

2. The further metabolism of pyruvate also yields ATP synthesis through **oxidative phosphorylation** (see Chapter 7).

3. Disruption of glycolysis causes disease and death due to the reliance of some tissues (RBCs and neurons, for example) on glucose metabolism for their energy needs.

4. The first steps in glycolysis result in the conversion of a six-carbon glucose molecule to **two three-carbon intermediates** (Figure 6–1).

5. Energy (ATP) is expended in the phosphorylation of intermediates in these reactions.

6. Many enzymes and intermediates of glycolysis also operate in gluconeogenesis.

B. Two key enzymes, **hexokinase** and **glucokinase,** catalyze the reaction of glucose with ATP to form **glucose 6-phosphate,** which becomes trapped in the cell and subject to metabolism.

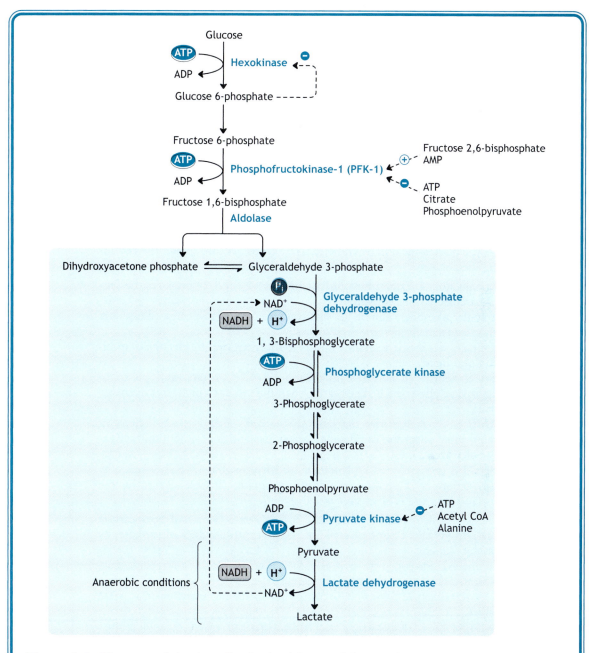

Figure 6–1. The steps of glycolysis. Feedback inhibition of glucose phosphorylation by hexokinase, inhibition of pyruvate kinase, and the main regulatory, rate-limiting step catalyzed by phosphofructokinase (PFK-1) are indicated. Pyruvate formation and substrate-level phosphorylation are the main outcomes of these reactions. Regeneration of NAD^+ occurs by reduction of pyruvate to lactate during anaerobic glycolysis.

1. The principal enzyme catalyzing this reaction, **hexokinase,** is found in all cells and has a **high affinity (low K_m)** for glucose.
   a. The high affinity of hexokinase for glucose means that even when glucose levels in the body are low, cells can efficiently take up glucose and obtain energy from it.
   b. Glucose 6-phosphate inhibits hexokinase, preventing cells from metabolizing excess glucose and harming other cells by reducing glucose available in the blood for metabolism.
2. **Glucokinase** is found in the liver and is responsible for dealing with the high levels of glucose available after a meal.
   a. Glucokinase has a **low affinity (high K_m)** for glucose but has a **high V_{max}.**
   b. Elevation of blood glucose levels after a meal stimulates the pancreas to secrete **insulin,** which among its many actions induces synthesis of glucokinase by the liver.
3. The **negative charge of glucose 6-phosphate** prevents it from diffusing across the plasma membrane and effectively **traps glucose inside the cell** for future metabolism.

C. The key regulatory enzyme **phosphofructokinase-1** (PFK-1) catalyzes the synthesis of **fructose 1,6-bisphosphate.**
   1. ATP is the phosphate donor for this reaction; two high-energy phosphates must be invested at the start of glycolysis.
   2. PFK-1 catalyzes this **irreversible** and **rate-limiting step in glycolysis** and is **highly regulated.**
   3. PFK-1 is subject to allosteric inhibition by ATP, citrate, and phosphoenolpyruvate, all of which are elevated when the cell has a high level of energy reserves.
      a. **AMP** is a very sensitive indicator of the cell's energy needs because of rapid interconversion of adenine nucleotides and is an important **activator of PFK-1.**
      b. PFK-1 is also activated by **fructose 2,6-bisphosphate,** which is made by the action of a second phosphofructokinase, PFK-2, using fructose 6-phosphate and ATP as substrates.

D. The six-carbon fructose 1,6-bisphosphate is then cleaved into two three-carbon molecules, **dihydroxyacetone phosphate** and **glyceraldehyde 3-phosphate,** by the action of aldolase.
   1. Interconversion between these three-carbon intermediates is a reversible reaction catalyzed by triose phosphate isomerase.
   2. Only glyceraldehyde 3-phosphate can go on to further metabolism to yield pyruvate.

E. In the second phase of glycolysis, two glyceraldehyde 3-phosphate molecules from glucose are converted to **pyruvate** in conjunction with several important **energy-generating reactions** (Figure 6–1).
   1. The formation of **1,3-bisphosphoglycerate** involves the synthesis of a high-energy phosphate bond as the aldehyde of glyceraldehyde 3-phosphate is oxidized to a carboxylic acid and then phosphorylated by reaction with inorganic phosphate.
   2. Formation of 1,3-bisphosphoglycerate is coupled to reduction of **NAD⁺** by transfer of two electrons and a proton to form **NADH + H⁺.**

F. Conversion of 1,3-bisphosphoglycerate to **3-phosphoglycerate** represents an oxidation coupled to synthesis of ATP from ADP.
   1. This reaction is catalyzed by phosphoglycerate kinase and is reversible.
   2. This is an example of **substrate-level phosphorylation,** ie, the creation of a high-energy phosphate bond through a chemical reaction rather than via oxidative phosphorylation (see Chapter 7).

G. Another key enzyme, **pyruvate kinase,** catalyzes the conversion of **phosphoenolpyruvate** to **pyruvate** and the formation of a second **ATP** in glycolysis.
   1. Pyruvate kinase is inhibited by compounds that are elevated when the cell has high energy reserves or molecules with potential for energy generation.
   2. High ATP levels inhibit pyruvate kinase.
   3. High amounts of **acetyl CoA** that can be converted to ATP through the tricarboxylic acid cycle (see Chapter 7) inhibit pyruvate kinase.
   4. High **alanine** levels inhibit pyruvate kinase; alanine can be converted to pyruvate (see Chapter 9).

PYRUVATE KINASE DEFICIENCY

- *Inherited deficiency of pyruvate kinase **impairs glycolysis** in all cells but has the most acute effect on RBCs.*
- *Anaerobic glycolysis is the only energy source available for maintenance of RBC viability, so the increased rate of erythrocyte death leads to **hemolytic anemia.***
- *Pyruvate kinase deficiency affects 1 in 10,000 people and is the most common inherited disorder of glycolysis.*
- *Most cases are due to decreased expression of pyruvate kinase activity, usually to **5–25% of normal levels;** complete loss of pyruvate kinase activity can cause embryonic death.*

III. Regeneration of NAD⁺

A. Regeneration of NAD$^+$ that was converted to NADH by electron transfer during glycolysis must occur in order for glycolysis to continue.

B. The mechanisms that maintain balance between NAD$^+$ usage and regeneration differ under **aerobic versus anaerobic** conditions in tissues.

C. In cells that are unable to transfer electrons to oxygen due to lack of mitochondria, eg, RBCs, or in vigorously exercising muscle cells (**anaerobic conditions**), NAD$^+$ is regenerated by further metabolism of pyruvate.
   1. Electrons are transferred from NADH to pyruvate by **lactate dehydrogenase,** forming NAD$^+$ and **lactate** (Figure 6–1).
   2. This reaction is reversible, and lactate can subsequently serve as an important source of carbons for gluconeogenesis in the liver.
   3. **RBCs lack mitochondria** and therefore depend on anaerobic glycolysis for energy needs.
   4. In muscle tissue under **hypoxic** conditions, the energy needs of the tissue may be partially supplied by anaerobic glycolysis.
      a. **Lactate build-up** during anaerobic glycolysis limits the extent to which muscle can obtain energy by this means.
      b. Accumulation of lactic acid causes a decrease in muscle cell pH.
      c. Decreased pH interferes with function of the contractile machinery of the muscle.

d. Elevated muscle lactate accounts for **fatigue** and **pain induced by strenuous exercise.**

D. In most cells, oxygen serves as the final acceptor of electrons removed during pyruvate synthesis (**aerobic conditions**).

   1. Electrons are removed from NADH and delivered to the mitochondrial electron transport chain where they ultimately are transferred to oxygen (see Chapter 7).

   2. Pyruvate is not consumed in this reaction and is available for further metabolism.

E. There are two shuttle mechanisms, the **malate-aspartate shuttle** and the **glycerol 3-phosphate shuttle,** that transport electrons to the inner mitochondrial matrix to be used in the electron transport chain.

   1. In the **malate-aspartate shuttle,** two electrons are transferred to form NADH in the inner mitochondrial matrix (Figure 6–2A).

      a. Oxaloacetate is reduced by reaction with NADH in the cytosol to form **malate** and regenerate NAD⁺ for glycolysis.

      b. Malate is then transported through the inner mitochondrial membrane by a transport protein.

      c. Oxaloacetate and NADH are then re-formed in the **mitochondrial matrix.**

      d. The electrons are passed from NADH to the electron transport chain for ATP biosynthesis in oxidative phosphorylation.

      e. Oxaloacetate is converted to **aspartate,** which is returned to the cytosol by transport across the inner mitochondrial membrane.

      f. Aspartate is converted to oxaloacetate, completing the cycle and allowing transport of more electrons to the mitochondria.

   2. The **glycerol 3-phosphate shuttle** is a second mechanism for transferring cytosolic electrons to the mitochondria (Figure 6–2B).

      a. Dihydroxyacetone phosphate is reduced by reaction with NADH to form **glycerol 3-phosphate** and NAD⁺.

      b. Two electrons are transferred from glycerol 3-phosphate to an **FAD complex,** which is imbedded in the inner mitochondrial membrane.

      c. FADH₂ is formed during this reaction and dihydroxyacetone phosphate is regenerated at the surface of the inner mitochondrial membrane.

      d. Electrons from FADH₂ are transferred to the electron transport chain.

   3. While the glycerol 3-phosphate shuttle appears to be less efficient than the malate-aspartate shuttle because fewer ATP molecules are synthesized (see Chapter 7), its advantage is that it enables the cell to transport electrons in the presence of high amounts of NADH.

F. The electrons that are generated from the first step in **ethanol metabolism** (catalyzed by **alcohol dehydrogenase**) are transported into the mitochondrion by these two shuttles.

LACTIC ACIDOSIS

- *Conditions that cause decreased oxygenation of tissues force **excessive** dependence on **anaerobic glycolysis** for energy production, with attendant **lactic acid buildup** in tissues and spillover into the blood.*

- *Convulsions, **shock,** uncontrolled hemorrhage or conditions that interfere with circulatory function can cause **lactic acidosis.***

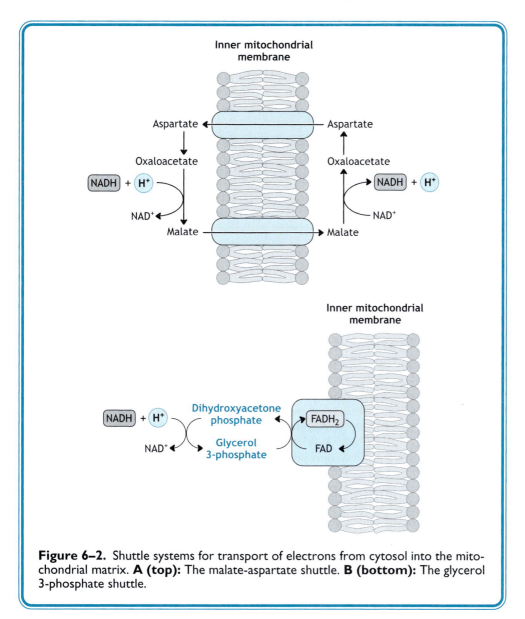

Figure 6–2. Shuttle systems for transport of electrons from cytosol into the mitochondrial matrix. **A (top):** The malate-aspartate shuttle. **B (bottom):** The glycerol 3-phosphate shuttle.

• **Metabolic acidosis,** *a potentially fatal condition, is characterized by nausea, vomiting, abdominal pain, lethargy, elevated heart rate, and irregular heart rhythm.*

• *Treatment of metabolic acidosis usually involves **intravenous sodium lactate** solution to normalize blood pH; the cause of the lactic acid overproduction should be determined and treated.*

    G. Energy yields from glycolysis depend on the system used to regenerate NAD⁺.

        1. Although ATP is consumed in the initial steps of glycolysis, it is generated in subsequent reactions, resulting in net ATP production.

        2. Under anaerobic conditions, glycolysis results in a net synthesis of only **two ATP** molecules for each molecule of glucose metabolized (Table 6–1).

Table 6–1. Energy yield in anaerobic glycolysis.

Enzyme Step	ATP Yield
Hexokinase	– 1
Phosphofructokinase-1	– 1
Phosphoglycerate kinase	+2
Pyruvate kinase	+2
Sum	**+2**

3. The energy yield resulting from glucose metabolism under aerobic conditions includes
   a. **Two ATP** molecules generated during anaerobic glycolysis.
   b. **Over 30 ATP** molecules are formed from the subsequent metabolism of pyruvate in the mitochondria (see Chapter 7).

IV. Pentose Phosphate Pathway

A. The **pentose phosphate pathway** (PPP), also called the **hexose monophosphate shunt,** is an alternate pathway of glucose metabolism that supplies the NADPH required by many biosynthetic pathways.
   1. The main purpose of the PPP is to **generate NADPH** to be used in **pathways for synthesis** of important molecules, eg, amino acids, lipids, and nucleotides.
   2. NADPH derived from the PPP is also important for **detoxification** of reactive oxygen species.
   3. The PPP also is responsible for **synthesis of ribose 5-phosphate** for nucleotide biosynthesis.

B. The PPP operates in two phases: an **oxidative phase** and a **nonoxidative phase.**
   1. In the **oxidative phase,** glucose 6-phosphate is metabolized by **glucose 6-phosphate dehydrogenase** (G6PD) to form 6-phosphogluconolactone (Figure 6–3).
      a. **NADP$^+$**, a coenzyme for this reaction, is **reduced to NADPH + H$^+$** in this reaction.
      b. G6PD catalyzes this **rate-limiting step** of the PPP and is inhibited by NADPH.
      c. Once this reaction occurs, 6-phosphogluconolactone is committed to the PPP.
      d. The oxidative steps of the PPP result in the formation of two molecules of NADPH, one of CO_2, and one molecule of ribulose 5-phosphate.
   2. The **nonoxidative phase** consists of a series of sugar-phosphate interconversions that result in the conversion of **ribulose 5-phosphate** to **ribose 5-phosphate** (Figure 6–3).
      a. Ribose 5-phosphate provides the ribose and deoxyribose sugars found in nucleotides.

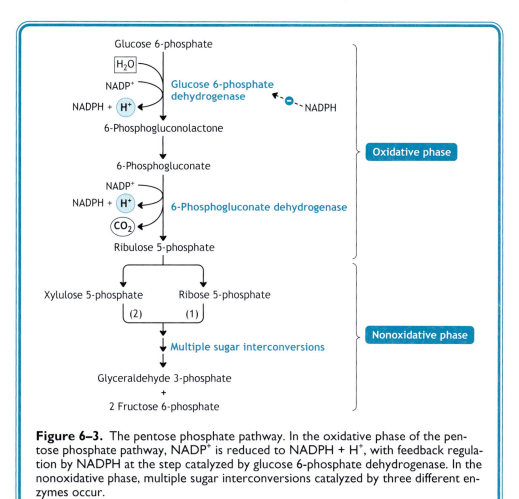

Figure 6–3. The pentose phosphate pathway. In the oxidative phase of the pentose phosphate pathway, $NADP^+$ is reduced to $NADPH + H^+$, with feedback regulation by NADPH at the step catalyzed by glucose 6-phosphate dehydrogenase. In the nonoxidative phase, multiple sugar interconversions catalyzed by three different enzymes occur.

   b. If adequate amounts of ribose are available through the diet and from cellular turnover of nucleotides, then an alternative branch of the PPP is used.
   (1) Two molecules of ribulose 5-phosphate are converted to xylulose 5-phosphate.
   (2) These intermediates then react with ribose 5-phosphate to form glycolytic intermediates that can be used for energy production.
   c. Thus, NADPH can be generated in the absence of net ribose production and the carbohydrate backbone can be used to make energy.

G6PD DEFICIENCY CAUSES SENSITIVITY TO OXIDANTS

- *G6PD deficiency is the most common genetic disease in the world, affecting over 400 million people, most of whom are men, because the gene is located on the X chromosome.*
- *Persons with G6PD deficiency are normally asymptomatic, but their RBCs are susceptible to **oxidative damage** because they have **impaired production of NADPH.***

– In affected persons, RBCs have a limited ability to detoxify **reactive oxygen species,** eg, hydrogen peroxide.
– Reactive oxygen species react with and denature cellular components, particularly hemoglobin, leading to premature RBC death and **hemolysis** unless they are reduced by glutathione, which is dependent on NADPH for its regeneration.
– The presence of **precipitates of oxidized, denatured hemoglobin (Heinz bodies)** helps distinguish the hemolytic anemia caused by of G6PD deficiency from that caused by pyruvate kinase deficiency.
– RBCs are especially sensitive to G6PD deficiency because the PPP is the only source of NADPH in these cells.
• Hemolytic anemia can be caused by eating foods (such as fava beans) or taking drugs (such as antimalarial agents or acetaminophen) that have oxidizing properties.
• The endemic presence of malaria in subequatorial Africa, where up to 25% of males are G6PD-deficient, is associated with G6PD deficiency because the malaria protozoan is less viable in RBCs with increased oxidative stress.

V. Key Enzymes Regulating Rate-Limiting Steps of Glucose Metabolism

A. Control of glucose degradation is accomplished by the regulation of key enzymes in the pathway (Table 6–2).

B. This regulation is related to the amount of energy stores in the cell and the availability of substrates for the generation of ATP.

VI. Glycogen Metabolism

A. Glycogen is the **storage form of glucose** mainly found in liver and muscle.

    1. Glycogen stores are regulated by a balance between glycogen synthesis (**glycogenesis**) and breakdown (**glycogenolysis**).

    2. Glycogen stores serve as an **easily mobilized source of glucose** to provide for the short-term needs of the body.

Table 6–2. Regulators of enzyme activity in glucose metabolism.

Enzyme	Activators	Inhibitors
Hexokinase		Glucose 6-phosphate
Phosphofructokinase-I	Fructose 6-phosphate AMP Fructose 2,6-bisphosphate Phosphoenolpyruvate	ATP Citrate
Pyruvate kinase		ATP Alanine Acetyl CoA
Glucose 6-phosphate dehydrogenase		NADPH

3. However, the glycogen stores of the body are insufficient to provide glucose during prolonged glucose deficit, ie, during fasting that lasts more than 24 hours.

B. Glycogenesis occurs in response to **stimulation by insulin** after ingestion of a meal that raises blood glucose levels.

   1. The first step of glycogenesis involves conversion of glucose 6-phosphate to **glucose 1-phosphate** by the action of **phosphoglucomutase** (Figure 6–4).
   2. Glucose 1-phosphate then is coupled with uridine diphosphate (UDP) to form **UDP-glucose**, the main **donor of glucosyl residues** during the construction of glycogen.
   3. **Glycogen synthase** catalyzes the addition of glucose from UDP-glucose to the end of a glycogen molecule, forming an α-1,4 linkage (Figure 6–4).

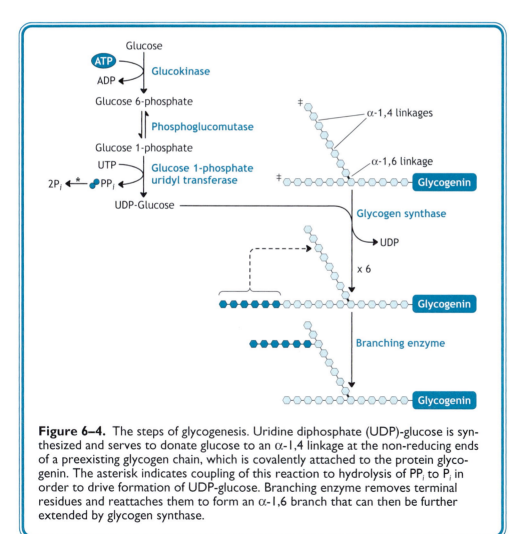

Figure 6–4. The steps of glycogenesis. Uridine diphosphate (UDP)-glucose is synthesized and serves to donate glucose to an α-1,4 linkage at the non-reducing ends of a preexisting glycogen chain, which is covalently attached to the protein glycogenin. The asterisk indicates coupling of this reaction to hydrolysis of PP_i to P_i in order to drive formation of UDP-glucose. Branching enzyme removes terminal residues and reattaches them to form an α-1,6 branch that can then be further extended by glycogen synthase.

a. This enzyme can only extend preexisting glycogen molecules.
   b. Glycogen synthase is only able to form α-1,4 glycosidic linkages as it extends the glycogen chain.

4. The glycogen recipient or acceptor is initially formed on **glycogenin,** a protein primer.

5. After a growing α-1,4 chain becomes approximately 11 glucose units in length, **branching enzyme** removes at least 6 units and adds them to another chain to form a branch.
   a. The branch is linked via an **α-1,6 glycosidic** bond to the main α-1,4 chain.
   b. Both the main chain and the new branch can then be extended by glycogen synthase.

C. As glucose is consumed by cellular metabolism, glycogen is degraded (**glycogenolysis**) to form free glucose in an effort to maintain relatively constant blood glucose levels.

1. Glycogen breakdown is catalyzed by **glycogen phosphorylase,** which removes the end glucose in the α-1,4 linkage from glycogen and combines it with inorganic phosphate to form glucose 1-phosphate (Figure 6–5).

2. Glucose 1-phosphate is then converted to glucose 6-phosphate by phosphoglucomutase.

3. When four glucose units remain on an α-1,4 branch of glycogen, then glycogen phosphorylase cannot remove any further glucose units.
   a. This impasse is overcome by **debranching enzyme,** which catalyzes transfer of an α-1,4-linked glucose trisaccharide to the end of another branch.
   b. Debranching enzyme then removes the remaining α-1,6-linked glucose as free glucose.

4. Glucose 6-phosphate can then be metabolized by glycolysis in the liver or muscle, or it can be dephosphorylated by the action of **glucose 6-phosphatase** mainly in the liver and released into the bloodstream for use by other tissues of the body.

GLYCOGEN STORAGE DISEASES

- *Deficiency of the enzymes of glycogen metabolism affects the ability of cells to store or use glycogen; as a result, regulation of blood glucose levels can be severely impaired during short-term fasting.*
- *Glycogen storage diseases produce **severe hypoglycemia,** even on an overnight fast, and are frequently diagnosed when the patient goes into **hypoglycemic shock** while sleeping.*
- *Untreated, glycogen storage diseases can lead to **mental retardation** or even death due to the energy loss in the brain consequent to low blood glucose levels.*
- *The most common glycogen storage disease, Type I or **von Gierke disease,** is a deficiency in glucose 6-phosphatase in which glycogen structures are normal; however, the liver is unable to dephosphorylate glucose 6-phosphate, and it remains trapped in the cell.*
- *Blood glucose levels in patients with von Gierke disease fall precipitously upon fasting, such as occurs overnight during sleep, so treatment is to eat meals often to prevent hypoglycemic coma.*

   D. Hormonal Regulation of Glycogen Metabolism
   1. As discussed in Chapter 5, the body regulates **blood glucose** concentration through the opposing actions of **insulin versus glucagon and epinephrine.**
      a. **Insulin** promotes **storage** of sugars, amino acids, and fats when they are available from dietary sources.

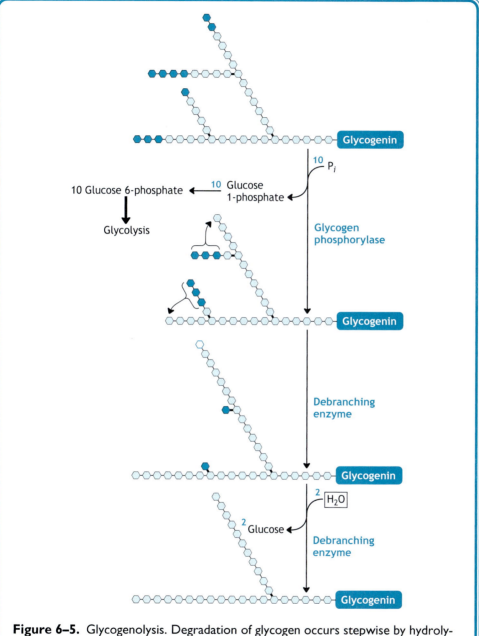

Figure 6–5. Glycogenolysis. Degradation of glycogen occurs stepwise by hydrolysis of one glucosyl unit at a time from the nonreducing ends by phosphorylase. The limit dextrin occurs as indicated in the second step when there are four glucosyl units remaining to a branch point. Once debranching enzyme has resolved the limit dextrin, degradation by phosphorylase can resume.

(1) Insulin simultaneously stimulates glycogen synthesis and inhibits glycogen breakdown.

(2) These combined **insulin** actions mediated mainly in liver and muscle promote a **net storage of glucose** for future needs.

b. Beginning about 4 hours after the last meal, blood glucose levels begin to fall, triggering an increase in **glucagon,** which promotes a changeover from net storage of fuel toward utilization.

(1) Glucagon stimulates glycogen breakdown and simultaneously inhibits glycogen synthesis in the liver.

(2) **Net glycogen breakdown** enables the liver to secrete glucose to provide for the energy needs of much of the body, **particularly the brain.**

c. **Epinephrine** actions provoke mobilization of various types of fuel reserves during times of **emergency energy need** (eg, stress).

(1) The actions of epinephrine and glucagon are very similar at the molecular level but occur mainly in different tissues.

(2) Thus, epinephrine causes a rapid response toward **net glycogen breakdown** to provide for the energy needs **of muscle.**

2. The hormonal mechanisms of action to regulate glycogenesis and glycogenolysis involve **reversible phosphorylation of the critical enzymes glycogen synthase and phosphorylase,** respectively (Figure 6–6).

a. **Glucagon** and **epinephrine** promote increased **phosphorylation** of glycogen synthase on serine residues, which **inactivates the enzyme.**

b. **Insulin** action, mediated by protein phosphatase I, causes **dephosphorylation** of glycogen synthase, which **activates the enzyme.**

c. Glycogen **phosphorylase is active** in degrading glycogen when it is **phosphorylated** on serine residues by a dedicated kinase.

d. Glucagon and epinephrine promote phosphorylation of the kinase, which in turn transfers the signal to phosphorylase.

e. Insulin action shuts down both the kinase and phosphorylase itself through activation of protein phosphatase I.

VII. Gluconeogenesis

A. Blood glucose levels must be maintained within a relatively **constant range to supply critical organs and tissues** (such as brain, RBCs, cornea, lens, kidney medulla, and testes), even when intake of dietary carbohydrates is low.

B. During a prolonged fast, glucose can be **synthesized from various precursors,** predominantly in the liver, by **gluconeogenesis.**

C. Because three of the reactions of the glycolytic pathway are irreversible, it is not possible to simply run glycosis in reverse to manufacture glucose.

1. The critical and irreversible **glycolytic steps that must be bypassed** follow:

a. Phosphoenolpyruvate (PEP) to pyruvate, catalyzed by pyruvate kinase.

b. Fructose 6-phosphate to fructose 1,6-bisphosphate, catalyzed by PFK-1.

c. Glucose to glucose 6-phosphate, catalyzed by hexokinase or glucokinase.

2. However, seven of the reactions of glycolysis are reversible and can be used for gluconeogenesis.

D. Conversion of **pyruvate to PEP** requires two enzyme-catalyzed steps.

1. Carboxylation of **pyruvate** to **oxaloacetate** occurs in the mitochondria (Figure 6–7).

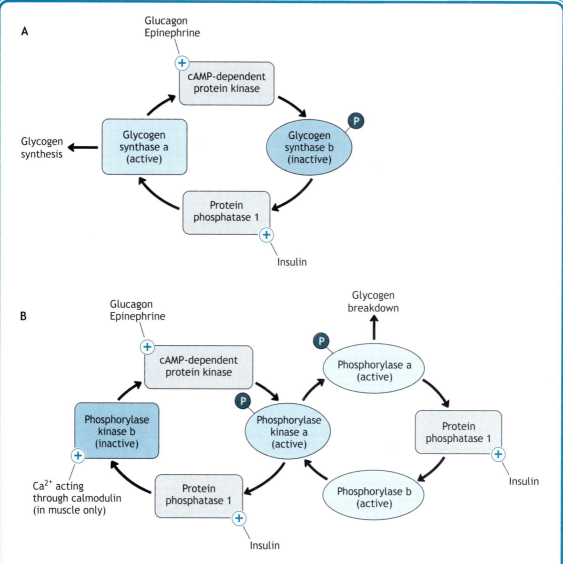

Figure 6–6. Hormonal regulation of glycogen metabolism. **A:** Glycogenesis. Activation of cyclic AMP (cAMP)–dependent protein kinase by the action of glucagon or epinephrine binding to their cell-surface receptors leads to phosphorylation and inactivation of glycogen synthase. Reactivation is catalyzed by protein phosphatase I, which is activated as a result of insulin binding to its cell-surface receptor. **B:** Glycogenolysis. The activity of glycogen phosphorylase is controlled by reversible phosphorylation, in a manner opposite to that of glycogen synthase. The effects of glucagon and epinephrine are still mediated by cAMP-dependent protein kinase, but through phosphorylase kinase, which itself is regulated by a phosphorylation-dephosphorylation cycle. Insulin action promotes dephosphorylation both of phosphorylase kinase and of phosphorylase itself, which inhibits glycogen breakdown.

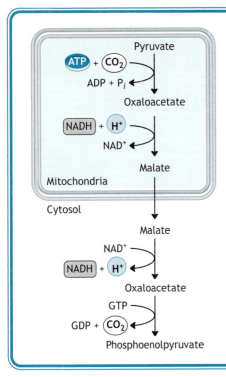

Figure 6–7. Conversion of mitochondrial pyruvate to cytosolic phosphoenolpyruvate to initiate gluconeogenesis. Oxaloacetate cannot pass across the inner mitochondrial membrane, so it is reduced to malate, which can do so.

    a. This important reaction is catalyzed by **pyruvate carboxylase.**
    b. ATP serves as an energy donor for the reaction of pyruvate with CO_2.
    c. Pyruvate carboxylase requires covalently bound **biotin** as a coenzyme to which CO_2 is temporarily attached during the transfer.
    d. Oxaloacetate can then enter the **tricarboxylic acid (TCA) cycle** to produce energy through oxidative phosphorylation or it may be used for gluconeogenesis.
  2. To initiate **gluconeogenesis,** oxaloacetate is reduced to **malate,** which is then transported to the **cytosol** in the **reverse of the malate shuttle.**
  3. Oxaloacetate is re-formed in the cytosol by oxidation of malate.
  4. Oxaloacetate is decarboxylated and simultaneously phosphorylated to **PEP.**
    a. This step requires the enzyme **PEP carboxykinase.**
    b. GTP hydrolysis provides the **energy** for this reaction and serves as the **phosphate donor.**
 E. The reactions of glycolysis converting fructose 1,6-bisphosphate to PEP are reversible, so that when glucose levels in the cell are low, equilibrium favors the conversion of PEP to fructose 1,6-bisphosphate (Figure 6–8).
 F. Conversion of fructose 1,6-bisphosphate to fructose-6-phosphate overcomes another of the irreversible steps of glycolysis and is catalyzed by **fructose 1,6-bisphosphatase** (Figure 6–8).
  1. This is an important **regulatory site** for gluconeogenesis.
  2. The reaction is allosterically **inhibited** by high concentrations of **AMP,** an indicator of an energy-deficient state of the cell.

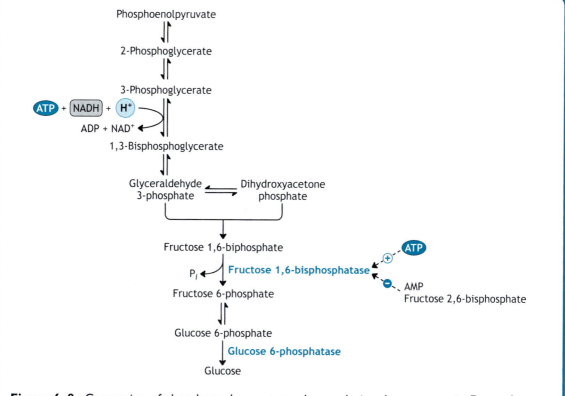

Figure 6–8. Conversion of phosphoenolpyruvate to glucose during gluconeogenesis. Except for the indicated enzymes that are needed to overcome irreversible steps of glycolysis, all other steps occur by the reverse reactions catalyzed by the same enzymes as those used in glycolysis.

    3. The enzyme is also inhibited by **fructose 2,6-bisphosphate,** which also functions as an allosteric activator of glycolysis.
    4. Conversely, the enzyme is subject to allosteric **activation by ATP.**
  G. Fructose 6-phosphate is isomerized to glucose 6-phosphate in a reversal of the glycolytic pathway.
  H. The initial irreversible step of glycolysis is bypassed by **glucose 6-phosphatase,** which catalyzes the **dephosphorylation of glucose 6-phosphate** to form glucose (Figure 6–8).
    1. This enzyme is mainly found in **liver** and **kidney,** the only two organs capable of releasing free glucose into the blood.
    2. A special transporter (**GLUT2**) in the membranes of these organs allows release of the glucose.

VIII. Metabolism of Galactose and Fructose
  A. The main dietary source of **galactose** is lactose.
    1. The disaccharide lactose is hydrolyzed by intestinal lactase.

2. Both of its component six-carbon sugars, glucose and galactose, then may be used for energy production.

B. Galactose and glucose are converted to uridine nucleotides and ultimately **interconverted** by a **4-epimerase,** which alters the orientation of the bonds at the 4 position of the molecule.

   1. In the cell, galactose is converted to galactose 1-phosphate by galactokinase with ATP as the phosphate donor.
   2. Galactose 1-phosphate and UDP-glucose react to form UDP-galactose and glucose 1-phosphate, as catalyzed by galactose 1-phosphate uridyltransferase.
   3. UDP-galactose can be converted to UDP-glucose by uridine diphosphogalactose 4-epimerase.
   4. The UDP-glucose can be used for glycogen biosynthesis.

GALACTOSEMIA

- *Galactosemia impairs metabolism of galactose to glucose, resulting in elevated blood galactose levels and **galactose accumulation** in tissues producing toxic effects in many organs.*
- *Patients may suffer **liver damage,** kidney failure, cataracts, mental retardation and, potentially, death in up to 75% of affected, untreated persons.*
- *Classic galactosemia is a rare, autosomal recessive disorder caused by deficiency of galactose 1-phosphate uridyltransferase.*
- *Once diagnosed, galactosemia can be treated by **restricting dietary galactose,** especially by excluding lactose from infant formulas.*

   C. **Fructose,** present in honey and in table sugar (sucrose) as a disaccharide with glucose, can comprise up to 60% of the sugar intake in a typical Western diet.

   1. In the muscle, **hexokinase** acts on fructose to form fructose 6-phosphate, which then enters glycolysis.
   2. In the liver, the enzyme **fructokinase** catalyzes the reaction of fructose with ATP to form fructose 1-phosphate.
      a. Fructose 1-phosphate is then cleaved to form dihydroxyacetone phosphate and D-glyceraldehyde by action of the enzyme **aldolase B.**
      b. D-glyceraldehyde is phosphorylated to form glyceraldehyde 3-phosphate, which can be metabolized in the glycolyic pathway.

DISORDERS OF FRUCTOSE METABOLISM

- ***Hereditary fructose intolerance** is due to **aldolase B deficiency** and is often diagnosed when babies are switched from formula or mother's milk to a diet containing fructose-based sweetening, such as **sucrose** or **honey.***
- *The inability to hydrolyze fructose 1-phosphate for further metabolism reduces availability of inorganic phosphate and decreases ATP levels.*
- ***Insufficient inorganic phosphate** (especially in the liver cells of affected persons who ingest a large amount of fructose) impairs gluconeogenesis, protein synthesis, and energy production by oxidative phosphorylation.*
- *Fructose intolerance causes vomiting, severe hypoglycemia, and kidney and liver damage that may lead to organ failure and death.*
- ***Essential fructosuria** is a benign, **asymptomatic** condition arising from deficiency of the enzyme fructokinase that causes a portion of fructose to be excreted in the urine.*

CLINICAL PROBLEMS

A 24-year-old man from Liberia is being treated for malaria with 30 mg daily of primaquine. After 4 days of treatment, he returns with the complaint that he "has no energy at all." Blood work indicates that he is severely anemic, and dense precipitates are present in otherwise normal-looking RBCs, which contain normal levels of adult hemoglobin. A week after suspending the primaquine treatment, he reports feeling better and his RBC count returns to normal.

1. What is the most likely explanation for this patient's reaction to treatment for his malaria?

   A. Sickle cell anemia

   B. Pyruvate dehydrogenase deficiency

   C. G6PD deficiency

   D. β-Thalassemia

   E. α-Thalassemia

A 9-month-old girl is suffering from vomiting, lethargy, and poor feeding behavior. Her mother reports that the symptoms began shortly after the baby was given a portion of a popsicle and mashed bananas by her grandparents. The baby's discomfort seemed to resolve after breastfeeding was resumed.

2. Which of the following is the most likely diagnosis?

   A. Pyruvate kinase deficiency

   B. G6PD deficiency

   C. Galactosemia

   D. Hereditary fructose intolerance

   E. Essential fructosuria

3. Which of the following organs or tissues does NOT need to be supplied with glucose for energy production during a prolonged fast?

   A. Lens

   B. Brain

   C. RBCs

   D. Liver

   E. Cornea

A woman returns from a yearlong trip abroad with her 2-week-old infant, whom she is breastfeeding. The child soon starts to exhibit lethargy, diarrhea, vomiting, jaundice, and an enlarged liver. The pediatrician prescribed a switch from breast milk to infant formula containing sucrose as the sole carbohydrate. The baby's symptoms resolve within a few days.

4. Which of the following was the most likely diagnosis?

 A. Pyruvate kinase deficiency

 B. G6PD deficiency

 C. Galactosemia

 D. Hereditary fructose intolerance

 E. Essential fructosuria

The drug metformin is useful in the treatment of patients with type 2 diabetes mellitus who are obese and whose hyperglycemia cannot be controlled by other agents. There are reports that some patients are predisposed to the toxic side effects of this drug, which include potentially fatal lactic acidosis.

5. Which of the following factors would likely increase the risk for this type of problem in a patient taking metformin?

 A. Cardiopulmonary insufficiency

 B. Inactivity

 C. Excessive weight

 D. Consumption of small amounts of alcohol

 E. Moderate exercise

6. Deficiency of which of the following enzymes would impair the body's ability to maintain blood glucose concentration during the first 24 hours of a prolonged fast?

 A. Glycogen synthase

 B. Phosphorylase

 C. Debranching enzyme

 D. PEP carboxykinase

 E. Fructose 1,6-bisphosphatase

ANSWERS

1. The answer is C. The response of this patient to taking primaquine, an oxidant, for his malaria is consistent with a diagnosis of G6PD deficiency. The presence of normally shaped RBCs argues against sickle cell anemia. The inclusions, Heinz bodies, in his RBCs are a hallmark of G6PD deficiency and distinguish it from pyruvate dehydrogenase deficiency. The possibility of a thalassemia is eliminated by the normal hemoglobin content of the RBCs. The onset of the anemia with the administration of a drug with known oxidative properties is an indicator of G6PD deficiency.

2. The answer is D. The main sugar in mother's milk is lactose. When the baby was given the fruit and the artificially sweetened popsicle, she was exposed to fructose for the first time and apparently is fructose intolerant. This diagnosis should be confirmed by genetic testing. Essential fructosuria is a benign condition that would not have produced

such severe symptoms. The symptoms are also consistent with galactosemia, but would be expected as a reaction to lactose intake.

3. The answer is D. Only the liver and kidneys can synthesize glucose by gluconeogenesis. All the other organs listed are dependent on provision of glucose from blood, either supplied by the diet or by gluconeogenesis in liver and the kidneys.

4. The answer is C. The patient's symptoms and course in response to a lactose-containing formula are consistent with a diagnosis of galactosemia. Pyruvate kinase deficiency and glucose 6-phosphate dehydrogenase deficiency would manifest as anemias and are seldom seen in an infant in the case of G6PD deficiency. G6PD deficiency is usually identified by the occurrence of a hypoglycemic coma following an overnight fast but is not normally accompanied by vomiting or diarrhea. While genetic screening tests required in most states identify newborns with galactosemia, these tests may not have been performed on a child born outside the United States.

5. The answer is A. Patients taking metformin are susceptible to lactic acidosis under conditions that lead to hypoxia, such as cardiopulmonary insufficiency. Metformin is contraindicated for people with preexisting heart or kidney disease, pregnant women, and those on severe diets. The drug should be discontinued before patients undergo surgery, which may involve fasting or lead to dehydration. In short, the drug exacerbates any condition that places demands on the anaerobic metabolism of glucose that could lead to excessive production or reduced utilization or clearance of lactic acid.

6. The answer is B. Glycogen is the main source of glucose during the first 24 hours of a prolonged fast. Lack of glycogen phosphorylase, the major enzyme responsible for hydrolysis of glycogen (glycogenolysis), would severely impair the ability of the liver to make glucose from glycogen. The only other enzyme listed that would have any potential effect would be debranching enzyme, which helps remove the α-1,6-linked branches from glycogen and is required for complete degradation of glycogen. The other enzymes are involved either in glycogen synthesis or gluconeogenesis and would not have any effect on glucose production from glycogen.

CHAPTER 7
THE TCA CYCLE AND OXIDATIVE PHOSPHORYLATION

I. Overview of the Tricarboxylic Acid (TCA) Cycle

A. The TCA cycle, also called the Krebs cycle, is the final destination for **metabolism of fuel molecules.**

    1. The **carbon skeletons** of carbohydrates, fatty acids, and amino acids are ultimately **converted to CO_2** and **H_2O** as the end products of their metabolism.

    2. Most fuel molecules enter the pathway as **acetyl coenzyme A (CoA),** but the carbon skeletons of the amino acids may also enter the TCA cycle at various points.

B. **Electrons** derived from the carbon skeletons are captured and transferred by the **electron transport chain** to oxygen, driving the **generation of ATP.**

    1. Most of the energy available to human cells is synthesized from the combined activity of the TCA cycle and the electron transport chain.

    2. Because molecular oxygen, **O_2, is the final electron acceptor** and ATP is formed by phosphorylation of ADP, the overall process is called **oxidative phosphorylation.**

C. The reactions of the TCA cycle occur entirely within the **mitochondrial matrix.**

II. Biosynthesis of Acetyl CoA

A. The main entry point for the TCA cycle is through generation of acetyl CoA by **oxidative decarboxylation of pyruvate.**

    1. Pyruvate derived from glycolysis or from catabolism of certain amino acids is transported from the cytoplasm into the mitochondrial matrix.

    2. A specialized pyruvate transporter is responsible for this step.

B. The **pyruvate dehydrogenase (PDH) complex,** which consists of multiple copies of three separate enzymes, catalyzes synthesis of acetyl CoA from pyruvate (Figure 7–1).

    1. PDH removes CO_2 and transfers the remaining acetyl group to the enzyme-bound coenzyme **thiamine pyrophosphate,**

    2. Dihydrolipoyl transacetylase transfers the acetyl CoA to its **lipoic acid** coenzyme with a reduction of the lipoic acid.

    3. Dihydrolipoyl dehydrogenase transfers electrons from lipoic acid to **NAD^+ to form NADH** and regenerate the oxidized form of lipoic acid.

    4. The overall reaction catalyzed by the PDH complex is shown below.

$$\text{Pyruvate} + NAD^+ + CoA \rightarrow \text{Acetyl CoA} + NADH + H^+ + CO_2$$

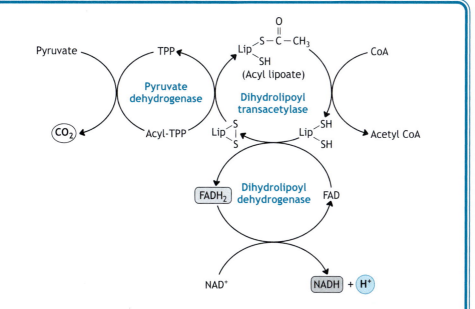

Figure 7–1. Conversion of pyruvate to acetyl CoA by the pyruvate dehydrogenase complex. The three enzymes, pyruvate dehydrogenase, dihydrolipoyl transacetylase, and dihydrolipoyl dehydrogenase, exist in a complex associated with the mitochondrial matrix. Each enzyme requires at least one coenzyme that participates in the reaction. TPP, thiamine pyrophosphate; Lip, lipoic acid; CoA, coenzyme A.

C. Regulation of PDH occurs through phosphorylation of the enzyme and by **allosteric regulation,** enabling a rapid response to changing energy needs of the cell or body.

1. **PDH kinase** inactivates PDH by phosphorylation of the enzyme.
   a. PDH kinase is activated by acetyl CoA, ATP, and NADH, all of which are indicators of high levels of cellular energy, thus promoting the inhibition of PDH.
   b. PDH kinase is inhibited by CoA, pyruvate, and by NAD^+, all found when cellular ATP levels are low.
2. **PDH phosphatase** removes the phosphate from PDH, returning the enzyme to its active form.
3. The unphosphorylated form of PDH also is subject to direct allosteric inhibition by NADH and acetyl CoA.

PDH DEFICIENCY

- *Deficiency in activity of the PDH complex disrupts mitochondrial fuel processing and may consequently cause neurodegenerative disease.*
  - *Loss of each of the PDH complex catalytic activities has been observed, with autosomal or X-linked (PDH) inheritance.*

- *Complete loss of PDH activity leads to neonatal death, while affected persons have detectable enzyme activity < 25% of normal.*
- *PDH deficiency may present from the prenatal period to early childhood, depending on the severity of the loss of enzyme activity, and there are no proven treatments for the condition.*
- *Symptoms of PDH deficiency include weakness, ataxia, and psychomotor retardation due to damage to the brain, which is the organ most reliant on the TCA cycle to supply its energy needs.*
- *Patients also suffer from lactic acidosis because the excess pyruvate that accumulates is converted to lactic acid.*
- *Other causes of PDH deficiency include a permanent activation of PDH kinase by its inhibitors or a loss of PDH phosphatase; in both cases, PDH is normal but remains in the phosphorylated or inhibited form regardless of the levels of its cellular regulators.*

III. Steps of the TCA cycle

    A. Acetyl CoA enters the TCA cycle by condensing with oxaloacetate to form **citrate** (Figure 7–2).

        1. This reaction is catalyzed by **citrate synthase.**

        2. Citrate rearranges to **isocitrate** in a reaction catalyzed by aconitase.

    B. Isocitrate dehydrogenase converts isocitrate to **α-ketoglutarate.**

        1. This is a dual reaction that combines decarboxylation to release CO_2 and oxidation, with capture of the electrons in NADH.

        2. Isocitrate dehydrogenase is the major regulatory enzyme of the TCA cycle.

    C. Conversion of α-ketoglutarate to **succinyl CoA,** CO_2, and NADH is catalyzed by the **α-ketoglutarate dehydrogenase** complex.

        1. This reaction again represents a combined **oxidation and decarboxylation.**

        2. By analogy to the PDH complex, the α-ketoglutarate dehydrogenase complex is made up of **three enzyme activities** with a similar array of activities and coenzyme requirements.

    D. Succinyl CoA is hydrolyzed to **succinate** and CoA in a reaction catalyzed by **succinyl CoA synthase.**

        1. This reaction involves simultaneous coupling of GDP and P_i to form GTP.

        2. This is another instance of **substrate-level phosphorylation.**

    E. Succinate is converted to **fumarate** with the transfer of electrons to FAD to form $FADH_2$, catalyzed by **succinate dehydrogense.**

    F. Fumarate undergoes hydration to **malate,** which is **converted to oxaloacetate,** completing the cycle.

        1. Another NADH is formed in the synthesis of oxaloacetate from malate.

        2. Oxaloacetate is then able to react with another acetyl CoA molecule to begin the cycle again.

    G. Oxidation of pyruvate yields CO_2, electrons, and GTP.

        1. The complete oxidation of one molecule of pyruvate can be described by the following equation:

$$\text{Pyruvate} + 4\ NAD^+ + FAD + GDP + P_i \rightarrow 3\ CO_2 + 4\ NADH + 4\ H^+ + FADH_2 + GTP$$

        2. One of the carbons of pyruvate is released as CO_2 during the formation of acetyl CoA.

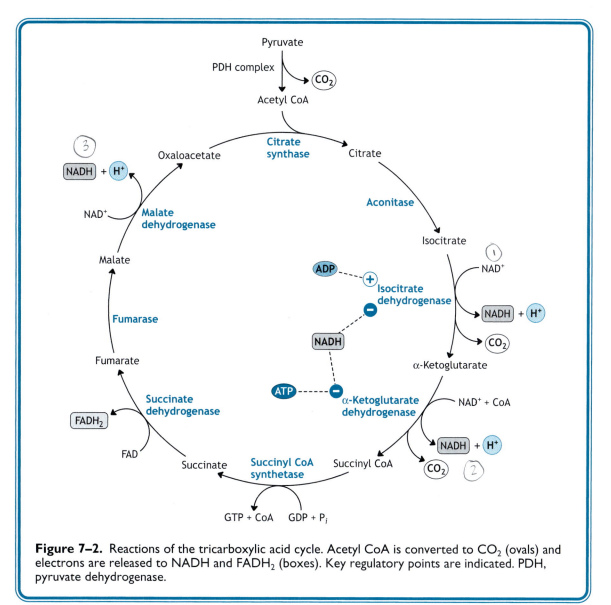

Figure 7–2. Reactions of the tricarboxylic acid cycle. Acetyl CoA is converted to CO_2 (ovals) and electrons are released to NADH and $FADH_2$ (boxes). Key regulatory points are indicated. PDH, pyruvate dehydrogenase.

3. During each turn of the TCA cycle, oxaloacetate is regenerated and metabolites of acetyl CoA are released.
   a. The two residual carbons of pyruvate are released as CO_2.
   b. **Five electron pairs** are extracted to enter the electron transport chain; ~~four~~ pairs are captured in NADH and one pair is captured in $FADH_2$.
4. Energy is also captured through substrate-level phosphorylation in the form of GTP synthesis.

THIAMINE DEFICIENCY

- *Thiamine pyrophosphate is an essential coenzyme for several critical metabolic enzymes—PDH, α-ketoglutarate dehydrogenase, and transketolase of the pentose phosphate pathway.*
- *Dietary deficiency of thiamine (**vitamin B₁**) results in an inability to synthesize thiamine pyrophosphate, and the pathophysiology arises from impaired glucose utilization, especially manifested in the nervous system.*
- *Thiamine deficiency is often seen as a nutritional disease in populations whose sole food source is polished rice, resulting in **beriberi.***
  - *In adults, symptoms include constipation, loss of appetite, nausea, peripheral neuropathy, weakness, muscle atrophy, and fatigue.*
  - *In nursing infants, the disease produces more profound symptoms, including tachycardia, convulsions and, potentially, death.*
- *Thiamine deficiencies are determined in the clinical laboratory by measuring the activity of **transketolase** in the RBC.*
- *Thiamine deficiency may also develop in alcoholics due to poor nutrition and poor absorption of thiamine in the gastrointestinal tract.*
- *In chronic alcoholics, thiamine deficiency may manifest as **Wernicke-Korsakoff syndrome,** which is characterized by a constellation of unusual neurologic disturbances, including amnesia, apathy, and nystagmus.*

ARSENIC TOXICITY

- *Arsenic can react irreversibly with the critical sulfhydryl groups of the coenzyme lipoic acid, which inactivates the coenzyme and thus inhibits the PDH complex and the α-ketoglutarate dehydrogenase complex.*
- *Symptoms of poisoning by arsenite (trivalent arsenic) include dermatitis and a variety of neurologic manifestations, including painful **paresthesias** (tingling and numbness in the extremities).*
- *Acute occupational exposures or direct ingestion cause **severe gastrointestinal distress** with diarrhea and vomiting, which may lead to dehydration, hypovolemic shock, and death.*

IV. Regulation of the TCA Cycle

    A. Availability of acetyl CoA from pyruvate is controlled by PDH activity, which is regulated by the concentration of NADH and the ADP/ATP ratio.

    B. The rate-limiting step of the TCA cycle is the synthesis of α-ketoglutarate from citrate, catalyzed by **isocitrate dehydrogenase** (Figure 7–2).

        1. Isocitrate dehydrogenase is allosterically inhibited by NADH, an indicator of the availability of high levels of energy.

        2. The enzyme is activated by ADP and Ca^{2+}, which signal a need for energy in the cell.

    C. Conversion of α-ketoglutarate to succinyl CoA, catalyzed by α-ketoglutarate dehydrogenase, is inhibited by NADH and ATP.

V. Role of the TCA Cycle in Metabolic Reactions

    A. Acetyl CoA and the TCA cycle intermediates are involved in many cellular reactions (Figure 7–3).

        1. Acetyl CoA is the precursor for fatty acid and sterol biosynthesis (see Chapter 8).

        2. The interconversion of α-ketoglutarate and glutamate are important for nitrogen metabolism.

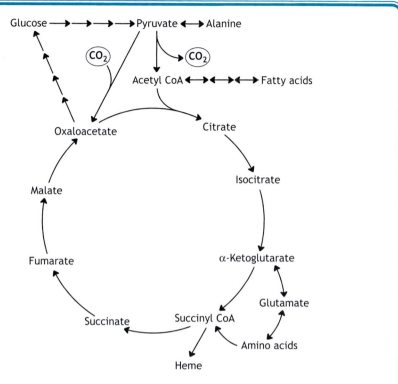

Figure 7–3. Interactions between metabolic pathways and the tricarboxylic acid cycle (TCA). Catabolic pathways feed carbon skeletons into the TCA cycle at various points to complete their metabolism. Acetyl CoA and several TCA cycle intermediates serve as precursors for synthesis of complex compounds.

3. The catalytic degradation of amino acids and pyrimidines yields pyruvate and several TCA cycle intermediates, which can then be metabolized in this way to yield energy.
4. Pyruvate and TCA cycle intermediates serve as precursors for the biosynthesis of amino acids (see Chapter 9).

VI. Synthesis of Oxaloacetate from Pyruvate

A. The ability to synthesize new oxaloacetate from pyruvate is essential to maintain activity of the TCA cycle for cell growth and for gluconeogenesis.
   1. **Pyruvate carboxylase** catalyzes the synthesis of oxaloacetate from pyruvate and CO_2.
   2. This reaction occurs within the mitochondria.
B. Oxaloacetate synthesis is also needed when mitochondria are formed during cell growth and division.
C. Oxaloacetate can also be converted to malate and transported to the cytoplasm for gluconeogenesis under fasting conditions (see Chapter 6).

PYRUVATE CARBOXYLASE DEFICIENCY

CLINICAL
CORRELATION

- *Deficiency of pyruvate carboxylase reduces oxaloacetate levels in the mitochondria, which limits TCA cycle activity with consequent impairment of many energy-requiring functions, eg, cell division.*
  - *Blockage of the TCA cycle causes accumulation of acetyl CoA, shunting to pyruvate and then lactate, which leads to **lactic acidosis.***
  - *Reduction of oxaloacetate synthesis also **impairs gluconeogenesis,** which compromises tissues dependent on glucose metabolism (such as the brain) during fasting.*
- *Pyruvate carboxylase deficiency is a rare disease that causes mental retardation and has led to death by age 5 in all known cases.*

VII. The Electron Transport Chain

A. The electrons released in glycolysis and transported into the mitochondria by shuttle mechanisms (see Chapter 6) and those derived from the TCA cycle are transferred to oxygen and combined with protons to form H_2O.

1. The **electron transport chain** is located in the **inner mitochondrial membrane** (Figure 7–4).

   a. The electron transport chain is organized into four complexes, each of which is composed of several integral membrane proteins and coenzymes capable of **reversible oxidation-reduction.**

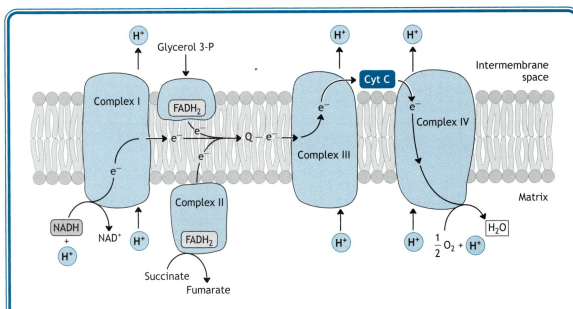

Figure 7–4. The electron transport chain. Electrons enter from NADH to complex I or succinate dehydrogenase, which is complex II. Electrons derived from glycolysis through the glycerol-3-phosphate shuttle, complex I, and complex II join at coenzyme Q and are transferred to oxygen as shown. As electrons pass through complexes I, III, and IV, protons are transported across the membrane, creating a pH gradient.

b. Each complex can accept electrons and then transfer them to other complexes through mediation of **mobile carriers, ubiquinone (coenzyme Q) and cytochrome c.**

    c. Electrons carried by NADH are transferred to **complex I.**

    d. Succinyl dehydrogenase of the TCA cycle is **complex II** with its FAD coenzyme, residing on the inner surface of the inner mitochondrial membrane.

  2. Electrons from both complex I and complex II are transferred to **ubiquinone,** a lipophilic compound residing in the membrane.

  3. Ubiquinone delivers electrons to **complex III,** which transfers them to complex IV via **cytochrome c.**

  4. Complex IV with its important **cytochrome a + a$_3$** catalyzes the **formation of water** from the electrons, protons, and oxygen.

VIII. Energy Capture During Electron Transport

A. As electrons pass through complexes I, III, and IV (but not complex II), **protons are transported** across the inner mitochondrial membrane from the matrix to the intermembrane space, creating a **pH gradient** that represents a form of stored energy.

B. The pH gradient is used to drive ATP synthesis by the movement of protons back to the matrix through a transmembrane protein complex, **or ATP synthase.**

  1. This mechanism was first described as the chemiosmotic theory of ATP generation, or the Mitchell hypothesis.

  2. As protons pass through a channel in the ATP synthase complex, ADP and P$_i$ are joined to form ATP.

C. ATP synthesized in the mitochondria is translocated to the cytoplasm by a co-transporter that simultaneously brings ADP into the mitochondria.

IX. Energy Yield of Oxidative Phosphorylation

A. The **ATP yield** from glucose metabolism via oxidative phosphorylation is approximately **34–36 ATP molecules per glucose molecule** (Table 7–1).

B. The calculated ATP yield is somewhat variable because glycolytic electrons transferred by the glycerol phosphate shuttle bypass complex I of the electron transport chain.

X. Inhibitors of ATP Generation

A. Transport inhibitors bind to one of the electron transport complexes and block the transfer of electrons to oxygen, thus interfering with the ability to create a proton gradient (Table 7–2).

B. The **ATP synthase inhibitor oligomycin** binds directly to the enzyme complex and plugs up the H$^+$ channel, which blocks ATP formation.

C. Uncoupling agents provide an alternate pathway to transfer protons back into the mitochondrial matrix, which dissipates the proton gradient and bypasses ATP formation by the ATPase.

  1. Thermogenin is a **natural uncoupler** found in the mitochondria of brown fat in hibernating animals and **infants.**

Table 7–1. Stoichiometry of ATP generation from one glucose molecule.[a]

	NADH	FADH$_2$	ATP
Cytoplasm			
Glucose → glucose 6-phosphate			-1
Fructose 6-phosphate → fructose 1,6-bisphosphate			-1
Glyceraldehyde 3-phosphate → glycerate 1,3-bisphosphate	+2		
Glycerate 1,3-bisphosphate → glycerate 3-phosphate			+2
Phosphoenolpyruvate → pyruvate			+2
Mitochondria			
Pyruvate → acetyl CoA	+2		

TCA cycle
  Oxidation of isocitrate, α-ketoglutarate, and malate
  Oxidation of succinate
  GDP → GTP
	+6	+2	+2

Oxidative Phosphorylation			
2 NADH from glycolysis			+6 (4)[b]
2 NADH from pyruvate → acetyl CoA			+6
6 NADH from TCA cycle			+18
2 FADH$_2$ from TCA cycle			+4
Total ATP			**+36(34)**

[a]Synthesis of NADH or FADH$_2$ and the subsequent conversion to ATP synthesis by oxidative phosphorylation is shown. It is assumed that approximately three molecules of ATP are made from the transfer of electrons from one NADH to oxygen and that two molecules of ATP are made from the electrons in FADH$_2$ going to oxygen.
[b]Six ATPs will be synthesized if the aspartate-malate shuttle is used to transfer NADH generated through glycolysis to NADH in the mitochondrial matrix; four molecules of ATP will be made if the glycerol phosphate shuttle delivers the electrons to ubiquinone in the inner mitochondrial membrane.
TCA, tricarboxylic acid.

    a. Thermogenin is a membrane protein that permits the organism to keep warm through metabolism without having to utilize ATP for movement.
    b. Under such conditions, up to 90% of ATP derived from fatty acid oxidation in these tissues is expended as heat.
  2. Chemical agents (such as **2,4-dinitrophenol**) that are able to bind a proton and be soluble in the lipid bilayer can also act as uncoupling agents.

Table 7–2. Inhibitors of ATP synthesis.

Inhibitor	Site of Action	Type
Rotenone	Complex I	Electron transport
Antimycin A	Complex III	Electron transport
Cyanide	Complex IV	Electron transport
Carbon monoxide	Complex IV	Electron transport
Azide	Complex IV	Electron transport
Thermogenin	Proton carrier	Uncoupler
2,4-Dinitrophenol	Proton carrier	Uncoupler
Oligomycin	ATP synthase	ATP synthase inhibitor

LEBER'S HEREDITARY OPTIC NEUROPATHY

- *Leber's hereditary optic neuropathy (LHON) is caused by a mutation of the ND1 gene encoding an element of **complex I** of the electron transport chain and other similar mutations.*
- *The pathophysiology of LHON arises from **impaired oxidative phosphorylation**, leading to **blindness** in many patients by early adulthood due to optic nerve death.*
- *The ND1 gene resides on the DNA of the mitochondria and is passed on to offspring by the egg cells of the mother, so there is no male-to-male transmission of LHON (see Chapter 13).*

CLINICAL PROBLEMS

A 2-year-old boy has a history of poor feeding and lethargy. He shows developmental delays and is in the fifth percentile for growth. His parents say that he has had no problems sleeping through the night but that he "just doesn't have any energy." A muscle biopsy and histologic examination show no apparent pathologic condition. Serum chemistry indicates severe lactic acidosis and hyperalaninemia. Supplementation of his diet with a B multivitamin does not alleviate his condition.

1. Which of the following is the most likely diagnosis?
   A. Pyruvate kinase deficiency
   B. PDH complex deficiency
   C. Pyruvate carboxylase deficiency
   D. Thiamine deficiency
   E. Niacin deficiency

A 2-month-old boy is brought to the emergency department in a coma after sleeping through the night and failing to awaken in the morning. He is given intravenous glucose and awakens. Serum levels of pyruvate, lactate, alanine, citrulline, and lysine are elevated, while aspartic acid levels are reduced. A muscle biopsy shows no abnormalities and vitamin supplementation is ineffective.

2. Which of the following is the most likely diagnosis?

    A. Pyruvate kinase deficiency

    B. PDH complex deficiency

    C. Pyruvate carboxylase deficiency

    D. Thiamine deficiency

    E. Niacin deficiency

A 55-year-old man complains of disorientation. He cannot remember where he was yesterday and appears confused. Upon examination he appears to be in poor health and admits to a "slight problem recently" with alcohol. After consultation with his daughter who accompanied him, it appears that alcohol abuse has been a severe problem for the past 35 years. Despite his confusion, his motor skills are normal when allowing for the general state of his health. However, he is subject to fits of rapid eye movements bilaterally.

3. What is the most likely cause of the patient's amnesia?

    A. Stroke

    B. Blunt force trauma to the head

    C. Wernicke-Korsakoff syndrome

    D. Hypoglycemia due to poor diet

    E. Alzheimer's disease or senile dementia

A 25-year-old man who has had problems with his eyesight has started to notice central vision loss. His older sister has similar problems, and his mother is a homemaker who is legally blind, although she told him that she used to be able to drive a car. He states he has no other medical problems. Consultation with an ophthalmologist indicates that his intraocular pressures are normal and that his lenses are clear. There is no sign of retinal bleeding. The patient is concerned that the same problem will develop in his children when they reach his age.

4. What is the problem with this patient?

    A. Stroke

    B. Leber's hereditary optic neuropathy (LHON)

    C. Macular degeneration

    D. Cataracts

    E. Glaucoma

A 7-year-old boy arrives at the emergency department asleep in his father's arms. The boy's mother explains that the boy spent the night throwing up and experiencing severe diarrhea. She is concerned about the vomiting and his inability to stay awake. History indicates the boy was healthy yesterday, but became ill at dinnertime after spending time playing in the basement of their apartment complex that afternoon. Further inquiry reveals that an exterminator had been hired to take care of a rat problem in the apartment, so she is worried that the boy may have been bitten by a rat. The boy is pale and not cyanotic. Chelation therapy is started for possible heavy metal poisoning, and poison control is notified.

5. An analysis of this patient's metabolism would likely indicate reduced activity of which of the following enzymes?

 A. PDH complex

 B. Pyruvate carboxylase

 C. Phosphofructokinase

 D. ATP synthase

 E. Citrate synthase

ANSWERS

1. The answer is B. While all of the listed conditions are consistent with lethargy and developmental defects, the lactic acidosis rules out pyruvate kinase deficiency. Thiamine and niacin deficiencies are unlikely due to the lack of effect of vitamin supplementation. Excess pyruvate is the source of the elevated alanine in the serum. The clinical findings are thus consistent with pyruvate carboxylase deficiency, which is associated with severe hypoglycemia due to fasting due to impaired gluconeogenesis.

2. The answer is C. Pyruvate kinase deficiency is ruled out by the elevated serum lactate levels. The coma is associated with a fasting hypoglycemia, which is indicative of pyruvate carboxylase deficiency. The elevated citrulline and lysine in the serum are due to a reduction of aspartic acid levels, which are caused by the reduced levels of oxaloacetate, the product of the pyruvate carboxylase reaction.

3. The answer is C. Poor diet and the scarring effects of long-term excessive alcohol ingestion on thiamine absorption in the intestine have led to thiamine deficiency and a related reduction of the activity of the PDH complex. The presence of chronic liver disease associated with long-term alcohol abuse reduces the ability to convert dietary thiamine to thiamine pyrophosphate, the active coenzyme of PDH. The long-term reduced energy metabolism in the brain caused by thiamine deficiency is thought to cause the neurologic damage leading to amnesia, which is due to irreversible cellular damage in the diencephalon. The normal motor skill assessment argues against stroke. While senile dementia and Alzheimer's disease may be present, they are less likely.

4. The answer is B. LHON often has an onset in early adulthood. It is a mitochondrial disorder usually resulting from a mutation in one of the proteins of the electron transport chain, particularly complex I, encoded by the mitochondrial genome; so there is no chance that the patient can pass the disorder to his children (see Chapter 13). Cataracts would have been detected as opacity in the lenses, and glaucoma would have been identified by an elevated intraocular pressure. Macular degeneration is also associated with central vision loss but is found mainly in patients over age 65.

5. The answer is A. This patient exhibits several signs of acute arsenic exposure, including the cholera-like gastrointestinal symptoms and probable dehydration. He may currently be in hypovolemic shock and beginning chelation therapy is the only recourse. Arsenic is a metabolic toxin because it inhibits enzymes that require lipoic acid as a coenzyme: the PDH complex, the α-ketoglutarate dehydrogenase complex, and transketolase of the pentose phosphate pathway.

CHAPTER 8
LIPID METABOLISM

I. Digestion and Absorption of Dietary Fats

A. Fats or **lipids** are **water-insoluble** and tend to coalesce into **droplets in water,** so a critical first step in processing dietary fats is **emulsification.**

   1. Emulsification breaks lipid droplets into smaller-sized structures, which **increases** their overall **surface area.**

   2. This process involves mixing (peristalsis) in the duodenum with **bile salts,** which act like **detergents** to dissipate lipid droplets.

   3. The increased contact area between water and lipids facilitates **interaction with digestive enzymes.**

B. Dietary lipids are processed by several **pancreatic lipases,** whose actions facilitate uptake by intestinal epithelial cells **(enterocytes).**

   1. **Triacylglycerols** are hydrolyzed by pancreatic **lipase** at their 1 and 3 positions.

      a. Lipase action cleaves triacylglycerols into two types of product: **free fatty acids (FFAs)** and **2-monoacylglycerols.**

      b. The drug **orlistat inhibits lipases** and thereby prevents uptake of many fats as a means of **treating obesity** in conjunction with a low-calorie diet.

   2. **Phospholipids** are hydrolyzed by **phospholipases,** which remove a fatty acid from carbon 2, leaving a **lysophospholipid,** which may be further processed or absorbed.

C. These products of lipid digestion combine to form **mixed micelles,** which are taken up efficiently by enterocytes.

   1. The mixed micelles contain predominantly FFAs, 2-monoacylglycerols, and unesterified cholesterol in addition to other fat-soluble compounds, such as the **fat-soluble vitamins A, D, E, and K.**

   2. After uptake, the micelles are dismantled and their components are modified for shipment to other organs.

      a. Fatty acids are **activated** to CoA esters by fatty acyl CoA synthetase.

      b. The **fatty acyl CoAs** are used to **rebuild triacylglycerols** using the 2-monoacylglycerol backbones and catalyzed by triacylglycerol synthase.

   3. **Cholesteryl esters** are synthesized by combining free cholesterol with a fatty acid.

LIPID MALABSORPTION DISORDERS

- *Fat malabsorption* can be caused by a variety of clinical conditions.
  - *Inflammatory conditions such as* **celiac disease** *can scar the intestine and cause villous atrophy, thereby reducing the surface area for fat digestion and absorption.*
  - *Individuals who have had surgical resection of portions of the intestine, eg, due to treatment of* **Crohn's disease,** *may also have impaired absorption of dietary fats.*
  - *Hepatobiliary disease, such as* **liver cancer** *or* **obstruction of the bile ducts,** *may lead to insufficient bile salt production or delivery, which reduces emulsification of fats.*
  - **Cystic fibrosis** *can obstruct pancreatic ducts due to mucous plugging and impaired secretion of pancreatic enzymes such as lipase and phospholipases, which decreases hydrolysis and uptake of triacylglycerols.*
- *A major symptom of fat malabsorption is* **steatorrhea,** *production of bulky, foul-smelling feces that float due to high fat content, which may be accompanied by diarrhea and abdominal pain, and if sustained for a period of days or weeks, lead to* **deficiencies of the fat-soluble vitamins.**

II. The Lipoproteins: Processing and Transport of Fats

A. Dietary fats are packaged by the enterocytes into **chylomicrons,** a very large type of lipid-protein complex or **lipoprotein,** for export to other organs.

  1. The triacylglycerols and cholesteryl esters form the **hydrophobic core** of the chylomicrons, which are coated with **surface phospholipids,** free cholesterol, and **apolipoprotein B-48.**

  2. Chylomicrons are discharged from the enterocytes by **exocytosis into lacteals,** which are lymphatic vessels that originate in the intestinal villi, drain into the cisternae chyli, and follow a course through the thoracic ducts to enter the bloodstream through the left subclavian vein.

B. The triacylglycerols of chylomicrons are degraded to FFAs and glycerol in many tissues, but especially in **skeletal muscle** and **adipose tissue.**

  1. Hydrolysis of triacylglycerols is catalyzed by **lipoprotein lipase,** a membrane-bound enzyme located on the **endothelium** lining the capillary beds of the muscle and adipose tissue.

  2. FFAs are then available for uptake by adipocytes or muscle cells.

  a. Within adipocytes, fatty acids can be **oxidized to yield energy** or re-esterified to glycerol for **storage** as triacylglycerols.

  b. Muscle cells can also utilize FFAs for energy.

  3. Fatty acids are transported in the blood **bound to albumin** for uptake and utilization by other tissues.

C. Most **plasma cholesterol** is esterified to fatty acids and is thus highly water-insoluble. These **cholesteryl esters** circulate in complexes with the lipoproteins.

D. The **lipoproteins** include chylomicrons, **HDLs,** intermediate-density lipoproteins (**IDLs**), **LDLs,** and **VLDLs,** which differ by size, density, and composition of proteins and lipids.

  1. Lipoproteins have a spherical **core of neutral lipids,** such as cholesteryl esters and triacylglycerols, which is coated with unesterifed cholesterol, phospholipids, and **apolipoproteins.**

  a. The apolipoproteins **mediate interaction** of the particles **with receptors and enzymes** involved in their metabolism.

b. The apolipoproteins specify the site of **peripheral uptake** of the lipoproteins, by mediating binding to receptors.
     2. The lipoproteins also have distinct structures and functions in the body.

E. **VLDLs** have **high triacylglycerol** content and are used to distribute fatty acids throughout the body.
     1. They are assembled in the liver and secreted into the bloodstream.
     2. The action of lipoprotein lipase lining the blood vessels degrades the triacylglycerols, releasing fatty acids locally for cellular uptake.
     3. In addition, triacylglycerols can be transferred to HDL particles transforming the VLDL into LDL.

F. **LDL particles,** the main carriers of **cholesterol** in the bloodstream, are taken up into cells by a receptor-mediated mechanism.
     1. The protein components of the LDL particles are degraded to amino acids.
     2. Cholesterol is then used by all cells as a component of the **plasma membrane** and other structures.
     3. Much of the LDL cholesterol is taken up by cells of the **liver,** where it is used to make **bile acids.**
     4. Many **steroidogenic tissues** synthesize **steroid hormones** from the cholesterol provided by LDL particles.

G. **HDL particles** have several functions, but among the most important is transport of excess cholesterol scavenged from the cell membranes back to the liver, a process called **reverse cholesterol transport.**
     1. HDL particles extract cholesterol from peripheral membranes and, after esterification of cholesterol to a fatty acid, the cholesteryl esters are delivered to the liver (to make bile salts) or steroidogenic tissues (precursor of steroids).
     2. In this way, **HDL particles** participate in **disposal of cholesterol,** and thus, a high HDL concentration is considered a protective factor against the development of cardiovascular disease.

III. Functions of Fatty Acids in Physiology

A. Fatty acids having at least 16 carbons (C16) play an important structural role as the major **components of cell membranes** (see Chapter 4).

B. Fatty acids comprise the principal **long-term fuel reserve** of the body in the form of triacylglycerols.
     1. These reserves are stored mainly in adipose and liver.
     2. Fatty acids stored as triacylglycerols are also generally ≥ C16.

C. In addition to fats that are made available from dietary sources, cells can synthesize many fatty acids.
     1. The most active organs in fatty acid synthesis are the **liver and the lactating mammary gland.**
     2. **Linoleic acid and linolenic acid** cannot be made in the body and are thus **essential.**
          a. **Linoleic acid** is a C18 fatty acid with two double bonds that is the precursor for synthesis of **arachidonic acid.**
          b. **Linolenic acid** is a C18 fatty acid with three double bonds that is the precursor for several other **omega-3 (ω-3) fatty acids.**

IV. Fatty Acid Synthesis

A. Fatty acids are constructed by stepwise addition of **two-carbon units** by a large multi-enzyme complex located in the cytoplasm of all cells.

  1. The two-carbon building blocks must be transported out of the mitochondria, where they exist in the form of **acetyl CoA.**

    a. The acetate unit is first transferred from acetyl CoA to oxaloacetate to form citrate by the enzyme **citrate synthase** (the first enzyme of the tricarboxylic acid [TCA] cycle; see Chapter 7).

    b. **Citrate** can pass across the inner mitochondrial membrane.

    c. Once in the cytoplasm, the acetate unit is transferred back to CoA by **ATP citrate lyase.**

  2. This pathway is active only when mitochondrial citrate and ATP concentrations are high, ie, when **high energy levels** are available.

    a. Thus, fatty acid synthesis is stimulated to allow storage of excess available two-carbon units as triacylglycerols.

    b. Fatty acid synthesis **requires large amounts of ATP and NADPH,** an energy investment that is largely recovered when the fatty acids are oxidized.

B. The **precursor** for donation of two-carbon units to build fatty acids is actually the three-carbon compound, **malonyl CoA.**

  1. Malonyl CoA is formed by carboxylation of acetyl CoA catalyzed by **acetyl CoA carboxylase.**

$$\text{Acetyl CoA} + CO_2 + ATP \rightarrow \text{Malonyl CoA} + ADP + P_i$$

  2. Formation of malonyl CoA is the **rate-limiting** and **principal regulatory step** of fatty acid synthesis.

    a. The enzyme is allosterically activated by citrate and is inhibited by long-chain fatty acyl CoA (end product inhibition).

    b. Acetyl CoA carboxylase is also regulated by reversible phosphorylation and dephosphorylation (Figure 8–1A).

      (1) **Glucagon** and **epinephrine** inactivate the pathway by promoting phosphorylation of the enzyme in order to **divert acetyl CoA toward energy generation** under conditions of low glucose and ATP levels.

      (2) **Insulin** action causes the enzyme to be dephosphorylated and therefore activated when blood glucose is elevated, in order to **stimulate storage of fuel as fat.**

    c. **Biotin** is a coenzyme for acetyl CoA carboxylase.

C. **Fatty acid synthase** is a large **multi-enzyme complex** that catalyzes the addition of two-carbon units in a **seven-step cycle** (Figure 8–2).

  1. During the reaction, acetate, or the growing fatty acyl chain is initially esterified to the sulfhydryl group of a cysteine residue of the enzyme.

  2. Malonate binds to the **phosphopanthotheine coenzyme** site and then the acetyl or acyl group is transferred to carbon two of malonate, with the loss of one malonyl carbon as CO_2.

  3. Subsequent reactions reduce the carbonyl group and reset the enzyme to accept the next two-carbon unit.

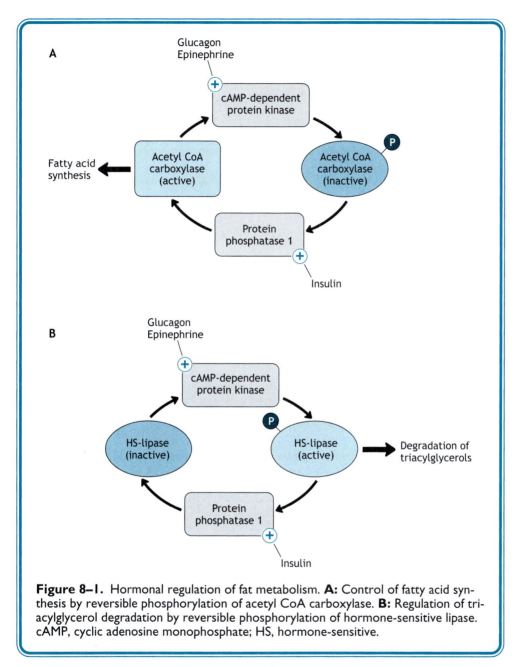

Figure 8–1. Hormonal regulation of fat metabolism. **A:** Control of fatty acid synthesis by reversible phosphorylation of acetyl CoA carboxylase. **B:** Regulation of triacylglycerol degradation by reversible phosphorylation of hormone-sensitive lipase. cAMP, cyclic adenosine monophosphate; HS, hormone-sensitive.

4. The reaction for each cycle indicates the high demand for ATP and also for **reducing equivalents** provided by NADPH, which are provided by the **pentose phosphate pathway** (see Chapter 6).

$$\text{Fatty Acyl(n) CoA} + \text{Malonyl CoA} + 2\text{NADPH} + 2\text{H}^+ \rightarrow \text{Fatty Acyl(n+2)}$$
$$\text{CoA} + \text{NADP}^+ + \text{CO}_2 + \text{H}_2\text{O}$$

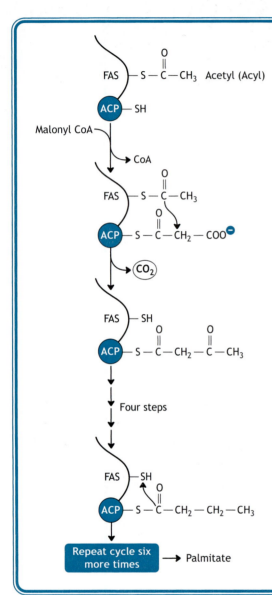

Figure 8–2. Pathway for synthesis of palmitate by the fatty acid synthase (FAS) complex. Schematic representation of a single cycle adding two carbons to the growing acyl chain. Formation of the initial acetyl thioester with a cysteine residue of the enzyme preceded the first step shown. Acyl carrier protein (ACP) is a component of the FAS complex that carries the malonate covalently attached to a sulfhydryl group on its phosphopantatheine coenzyme (-SH in the scheme).

    5. The **ultimate product** of seven cycles of these reactions is the fully saturated, C16 fatty acid **palmitate.**

  D. Additions to and **modifications of palmitate** allow synthesis of many structurally distinct fatty acids.

    1. Elongation of palmitate occurs by addition of further acetate units in the endoplasmic reticulum and mitochondria.

    2. Desaturation or the creation of double bonds for synthesis of **unsaturated fats** is performed by mixed-function oxidases in the endoplasmic reticulum.

3. **Storage as triacylglycerols** requires activation of the fatty acid by **conversion to acyl CoA** with glycerol 3-phosphate as the precursor for the glycerol backbone.

V. Fatty Acid Oxidation

A. Mobilization of fat stores allows fats to be burned to **produce energy** via **fatty acid oxidation.**

   1. The initial step to release fatty acids is **triacylglycerol hydrolysis** catalyzed by **hormone-sensitive (HS) lipase.**
      a. As its name implies, the enzyme is regulated via hormonally controlled cycles of phosphorylation and dephosphorylation (Figure 8–1B).
      b. **Glucagon** and **epinephrine stimulate lipase** activity in order to provide fatty acids and glycerol for use as fuels, while **insulin inhibits** lipase activity as it stimulates storage of fatty acids.
   2. The **glycerol backbone** derived from lipase-mediated triacylglycerol breakdown is released into the bloodstream and taken up by the liver.
      a. Glycerol is phosphorylated on its 3 position.
      b. **Glycerol 3-phosphate** can then enter **glycolysis** or **gluconeogenesis** (see Chapter 6).

B. Before oxidation can begin, the fatty acids must again be **activated by esterification with CoA.**

$$\text{Fatty Acid} + \text{CoA} + \text{ATP} \rightarrow \text{Fatty Acyl CoA} + \text{AMP} + \text{PP}_i$$

   1. Acyl CoA synthase combines the FFA with CoA.
   2. This reaction requires **energy input** provided by ATP hydrolysis.

C. **Long-chain fatty acids (LCFAs),** which have carbon chain lengths of 12–22 units (C12–C22), must be transported into the mitochondrial matrix where the enzymes responsible for their oxidation are located. This is accomplished by the **carnitine shuttle** (Figure 8–3).

   1. LCFAs are reversibly transesterified from CoA to **carnitine,** an amino acid derivative that serves as the carrier.
      a. Two enzymes, **carnitine palmitoyltransferases I and II (CPT-I and CPT-II),** located in the outer and inner mitochondrial membranes, catalyze this set of reactions.
      b. A **translocase** transporter binds acyl-carnitine and mediates its transport across the main barrier, the inner mitochondrial membrane.
   2. **Malonyl CoA,** an indicator that fatty acid synthesis is active in the cytoplasm, is an **inhibitor of CPT-I.**

CARNITINE DEFICIENCY LEADS TO MYOPATHY AND ENCEPHALOPATHY

- *Carnitine deficiency* leads to impaired carnitine shuttle activity; the resulting *decreased LCFA metabolism* and accumulation of LCFAs in tissues and wasting of acyl-carnitine in urine can produce cardiomyopathy, skeletal muscle myopathy, encephalopathy, and impaired liver function.
- *There are two recognized types of carnitine deficiency—primary and secondary.*
- *Primary carnitine deficiency* arises from inherited *deficiency of CPT-I or CPT-II,* both of which are rare disorders showing autosomal recessive inheritance.

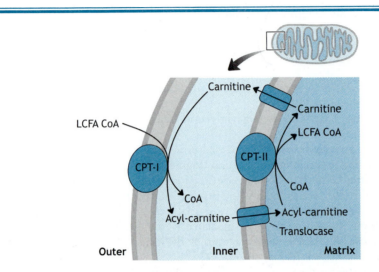

Figure 8–3. The carnitine shuttle. A long-chain fatty acyl CoA (LCFA CoA) can diffuse across the outer mitochondrial membrane but must be carried across the inner membrane as acyl-carnitine. The active sites of CPT-I and CPT-II are oriented toward the interiors of their respective membranes. CPT, carnitine palmitoyltransferase.

– *CPT-I deficiency produces a fasting hypoglycemia due to impaired liver function as a consequence of the inability to utilize LCFAs as fuel.*
– *CPT-II deficiency is more common and mainly manifests as muscle weakness, myoglobinemia, and myoglobinuria upon exercise; severe cases lead to hyperketotic hypoglycemia, hyperammonemia, and death.*
– *Both these disorders are treated by **avoidance of fasting, dietary restriction of LCFAs,** and **carnitine supplementation;** the objective is to stimulate whatever carnitine shuttle activity is present.*
• *Carnitine deficiency may also be **secondary** to a variety of conditions.*
– ***Impaired carnitine synthesis** due to liver disease.*
– ***Disorders of β-oxidation.***
– *Malnutrition due to consumption of some vegetarian diets.*
– *Depletion by hemodialysis.*
– *Increased demand due to illness, trauma, or pregnancy.*

    D. The reactions of **β-oxidation** cleave fatty acids in a series of cycles, each of which shortens the chain by **two carbons** (Figure 8–4).

        1. The initial step in each cycle of β-oxidation is catalyzed by one of several **acyl CoA dehydrogenases,** which are selective for fatty acids of different chain length.

        2. There are two oxidative steps at each cycle, producing **one FADH$_2$ and one NADH.**

        3. The products at the end of each cycle are **acetyl CoA** plus the fatty acyl CoA shortened by two carbons.

        4. The carbons of even-chained fatty acids end up producing **acetyl CoA in the final step.**

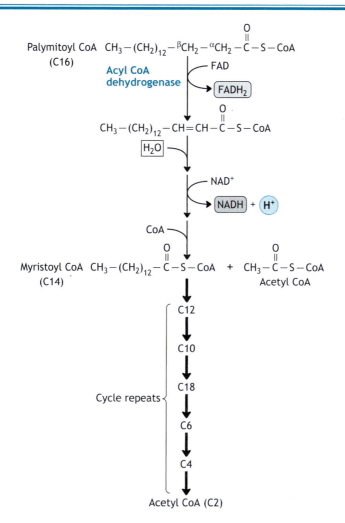

Figure 8–4. β-Oxidation of palmitate. Oxidation of an even-numbered, saturated fatty acid involves repetitive cleavage at the β carbon of the acyl chain. Removal of two-carbon units occurs in a cycle of four steps initiated by one of the acyl CoA dehydrogenases. Acetyl CoA is produced at each cycle until all that remains of the acyl CoA is acetyl CoA itself.

5. The **reaction at each cycle** (below) hints at the energy potential for β-oxidation of a fatty acid.

Fatty Acyl(n) CoA + FAD + NAD⁺ + CoA + H₂O → Fatty Acyl(n-2)
        CoA + FADH₂ + NADH + H⁺ + Acetyl CoA

a. Passage of the electrons from one FADH₂ and one NADH through the electron transport chain yields five ATP.

b. Extraction of energy from the electrons of each molecule of acetyl CoA via the TCA cycle and the electron transport chain would produce 11 more ATP.

c. One substrate phosphorylation reaction in the TCA cycle yields one ATP.

d. Thus, each two-carbon unit of a saturated fatty acid yields as much as 17 ATP.

e. Burning of a single molecule of **palmitate yields 131 ATP,** with a **net of 129 ATP** when the investment of ATP in the activation step is subtracted.

MCAD DEFICIENCY

- *Medium-chain fatty acyl CoA dehydrogenase (MCAD) deficiency* impairs metabolism of medium-chain (C6–C12) fatty acids.
  - The C6–C12 fatty acids and their esters accumulate in tissues to cause **toxicity.**
  - Spillover of C6–C10 acylcarnitine species into the blood provides for very specific diagnosis of MCAD.
- Children afflicted with MCAD deficiency experience muscle weakness, lethargy, **fasting hypoglycemia,** and **hyperammonemia,** which may lead to seizures, coma and, potentially, brain damage and death.
- MCAD deficiency is inherited in an autosomal recessive manner with an incidence of 1 in 8500 in the United States.
- MCAD deficiency is more common than SCAD deficiency, which impairs oxidation of short-chain (< C6) fatty acids, or LCHAD deficiency, which impairs oxidation of long-chain (C12–C22) fatty acids.
- Principal treatments of MCAD deficiency are to **avoid fasting** (even overnight), to **supplement with carnitine,** and to manage infections aggressively.

E. Oxidation of **odd-chain fatty acids** requires some specialized reactions.

1. The reactions of β-oxidation yield acetyl CoA molecules at each cycle as usual, leaving the three-carbon **propionyl CoA as a remnant.**

2. Propionyl CoA is further metabolized in a three-step process to **succinyl CoA,** in which methylmalonyl CoA is an intermediate.

a. Succinyl CoA can then enter the TCA cycle for further metabolism.

b. The enzyme **methylmalonyl CoA mutase** is one of only three enzymes of the body that require **vitamin B$_{12}$** as a coenzyme.

c. Excretion of propionate and **methylmalonate in urine** is a diagnostic hallmark of **vitamin B$_{12}$ deficiency.**

F. Oxidation of very long-chain fatty acids (**VLCFAs**), ie, fatty acids having >22 carbons, requires special enzymes located in the **peroxisome.**

1. A peroxisomal dehydrogenase initiates the β-oxidation reactions that **shorten the chain to ~18 carbons or less,** at which point the fatty acyl CoA is transferred to mitochondria for complete degradation by β-oxidation.

2. Dehydrogenation in the peroxisome produces **FADH$_2$.**

3. In order to sustain the pathway, FADH$_2$ must be reoxidized to FAD.

a. This is accomplished by reduction of molecular oxygen to **hydrogen peroxide, H$_2$O$_2$.**

b. Peroxide is then reduced to water by peroxisomal **catalase.**

G. Unsaturated fatty acids (ie, those having double bonds) can be metabolized through β-oxidation, but this process requires **additional enzymes.**

1. When a double bond appears near the carboxyl carbon of the partially degraded fatty acyl CoA, several isomerases and reductases modify the structure to allow continued β-oxidation.

2. Because they contain fewer electrons within their structures, **unsaturated fatty acids yield less energy** than corresponding saturated fatty acids in β-oxidation.

ZELLWEGER SYNDROME

- *Zellweger syndrome is a lipid storage disorder caused by **impaired peroxisome biogenesis** due to deficiency or functional defect of one of eleven proteins involved in the complex mechanism of peroxisomal matrix protein import and assembly of the organelle.*
  - *These defects suppress many peroxisomal functions, including **impaired oxidation of VLCFAs.***
  - *One of the genes responsible for this disorder, PEX5, encodes the import receptor itself.*
- *The cells have **absent** or **undersized peroxisomes** with accumulation of VLCFAs, which is especially marked in the liver, kidneys, and nervous tissue.*
- *Patients exhibit a broad spectrum of abnormalities, including liver and kidney dysfunction with **hepatomegaly,** high levels of copper and iron in the blood, **severe neurologic defects,** and skeletal malformations.*
  - *Such patients have a high incidence of perinatal mortality and rarely survive beyond 1 year.*
  - *The condition is of variable severity, but most forms are inherited in an autosomal recessive manner.*

X-LINKED ADRENOLEUKODYSTROPHY

- *X-linked adrenoleukodystrophy (X-ALD) is a progressive, **inherited neurologic disorder** arising from a **defect in peroxisomal VLCFA oxidation.***
  - *The gene for X-ALD encodes a peroxisomal membrane protein whose function is required for VLCFA oxidation, so VLCFAs accumulate in tissues and spill over into plasma and urine.*
  - *X-ALD is rare, with an incidence of 1 in 20,000–40,000.*
- *Symptoms arise in boys at about 4–8 years of age, manifested initially as dementia accompanied in most cases by adrenal insufficiency.*
  - *The most severely affected patients may end up in a **persistent vegetative state.***
  - *In some patients, milder symptoms develop, starting in the second decade, and include progressive **paraparesis** (weakness) in the lower extremities.*
- *MRI indicates a **severe reduction in cerebral myelin,** which likely accounts for the central neuropathy.*
- *VLCFAs arise from both dietary and endogenous synthetic sources, so treatment is mainly supportive.*
  - *Feeding a 4:1 mixture of glyceryl trioleate and glyceryl trierucate (**Lorenzo's Oil**) can reduce plasma VLCFA levels, but it is unclear whether this treatment can reverse demyelination.*
  - *Lovastatin and 4-phenylbutyrate are being tested as new therapeutic approaches to stimulate VLCFA metabolism.*

VI. Metabolism of Ketone Bodies

  A. Ketone body synthesis (ketogenesis) occurs only in the mitochondria of **liver** cells when acetyl CoA levels exceed the needs of the organ for use in energy production.

   1. Acetyl CoA is the precursor for all three ketone bodies, **acetoacetate, 3-hydroxybutyrate, and acetone.**

   2. Only acetoacetate and 3-hydroxybutyrate can be used as fuel by peripheral tissues.

    a. These compounds are **soluble in blood** and thus do not require lipoprotein carriers for transport to other tissues.

    b. The ketone bodies are **converted back to acetyl CoA** after uptake to be used **for energy** production in extrahepatic tissues.

c. Even the **brain can adapt** to use them as an energy source during long-term fasting.

  3. **Acetone** is a byproduct of acetoacetate decarboxylation and cannot be used as a fuel but is instead **expired via the lungs.**

B. Ketone body synthesis is active mainly during **starvation,** times of **intensive mobilization of fat** reserves by the adipose tissue.

  1. **High acetyl CoA** levels from β-oxidation of fatty acids in liver cells inhibit the pyruvate dehydrogenase complex and activate pyruvate carboxylase, which increases oxaloacetate synthesis.

  2. This shunts **oxaloacetate toward gluconeogenesis** and leaves acetyl CoA available for formation of ketone bodies.

  3. The pathway is initiated by condensation of two molecules of acetyl CoA to form acetoacetyl CoA (Figure 8–5A).

  4. Synthesis of hydroxymethylglutaryl CoA (**HMG CoA**) by condensation of acetoacetyl CoA with acetyl CoA is catalyzed by **HMG CoA synthase** and is the **rate-limiting step** of the pathway.

  5. Cleavage of HMG CoA yields **acetoacetate,** followed by reduction to **3-hydroxybutyrate,** which thus carries more energy than acetoacetate.

C. Utilization of ketone bodies by the extrahepatic tissues requires the activity of the enzyme **thiophorase** (Figure 8–5B).

  1. Conversion of 3-hydroxybutyrate to acetoacetate is necessary as a first step in its metabolism.

  2. Thiophorase then catalyzes transfer of CoA to acetoacetate to produce acetoacetyl CoA.

    a. Succinyl CoA is the donor for this **transesterification** reaction.

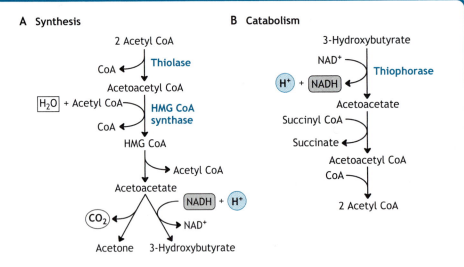

Figure 8–5. Pathways for metabolism of ketone bodies. **A:** Ketone body synthesis by the liver. **B:** Catabolism by conversion to acetyl CoA. Only organs that express thiophorase can utilize ketone bodies for energy.

b. Acetoacetyl CoA is then split into two molecules of acetyl CoA, which can enter the **TCA cycle for fuel.**

    c. The liver does not contain thiophorase, so it cannot use ketone bodies as fuel.

DIABETIC KETOACIDOSIS

- *Extremely **low insulin** levels in a person with uncontrolled type 1 diabetes mellitus produce **acidemia** and **aciduria** due to **high concentrations of ketone bodies,** which are acids and contribute to the decreased pH.*
  - *The condition is exacerbated by an accompanying **hyperglycemia** and unopposed glucagon action.*
  - *Dysfunction of fat metabolism is caused by the low insulin/glucagon ratio, which stimulates fat mobilization by adipose tissue, flooding the liver with fatty acids and raising intracellular acetyl CoA levels.*
  - ***Excess acetyl CoA** in the liver depletes NAD$^+$, and the high concentration of NADH blocks the TCA cycle.*
  - *This shunts acetyl CoA toward ketone body synthesis, which becomes excessive.*
- *These effects lead to major clinical manifestations, including nausea, vomiting, dehydration, electrolyte imbalance, loss of consciousness and, potentially, coma and death.*
- *A characteristic sign of this condition is a **fruity odor on the breath** due to expiration of large amounts of acetone.*

VII. Cholesterol Metabolism

A. Synthesis of **cholesterol** occurs in the cytoplasm of most tissues, but the liver, intestine, adrenal cortex, and steroidogenic reproductive tissues are the most active.

    1. Acetate, via acetyl CoA, is the initial precursor for cholesterol synthesis, leading in two steps to **HMG CoA.**

    2. Conversion of **HMG CoA to mevalonic acid** is catalyzed by the key regulatory enzyme, **HMG CoA reductase.**

        a. This is the **rate-limiting step** of cholesterol synthesis.

        b. HMG CoA reductase is **heavily regulated** by several mechanisms.

            (1) Expression of the HMG CoA reductase gene is controlled by a **sterol-dependent transcription factor,** which increases enzyme synthesis in response to low cholesterol levels.

            (2) **Insulin** up-regulates the gene and **glucagon** down-regulates it (Figure 8–6).

            (3) Enzyme activity is controlled by reversible phosphorylation/dephosphorylation in response to **AMP,** ie, cholesterol synthesis is suppressed when energy levels are low.

        c. The **statin drugs,** such as lovastatin, atorvastatin, and mevastatin, suppress endogenous cholesterol synthesis by **competitive inhibition of HMG CoA reductase,** and thereby act to **decrease LDL cholesterol.**

    3. **Mevalonic acid** is then modified by phosphorylation and decarboxylation, and several molecules of it are condensed to form cholesterol in a complex series of eight reactions.

B. **Bile salts** are synthesized by the **liver** with cholesterol as the starting material.

    1. Hydroxylation, shortening of the hydrocarbon chain, and addition of a carboxyl group convert cholesterol in a complex series of reactions to the **bile acids, cholic acid, and chenodeoxycholic acid.**

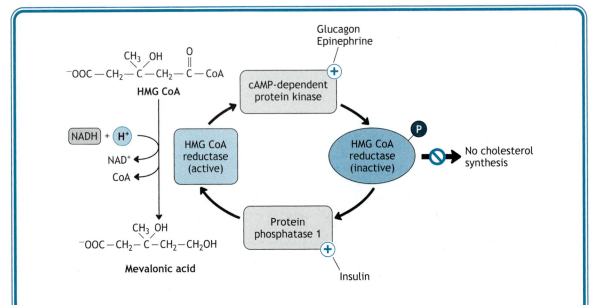

Figure 8–6. Hormonal regulation of cholesterol synthesis by reversible phosphorylation of HMG CoA reductase. Availability of mevalonic acid as the fundamental building block of the sterol ring system controls flux through the pathway that follows. cAMP, cyclic adenosine monophosphate; HMG CoA, hydroxymethylglutaryl CoA.

2. Subsequent **conjugation** of these acids **with glycine or taurine** forms the various **bile salts,** which have enhanced amphipathic character and are very effective **detergents.**
   a. Combination with glycine produces the common bile salts, **glycholic** and **glycochenodeoxycholic acids.**
   b. Conjugation with taurine, a derivative of cysteine, creates taurocholic and taurochenodeoxycholic acids.
3. The bile salts are either secreted directly into the duodenum or **stored in the gallbladder** for use in **emulsifying dietary fats** during digestion.
4. **Disposal in bile** either as bile salts or as cholesterol itself is the body's main mechanism for **cholesterol excretion.**

CHOLESTEROL GALLSTONE DISEASE

- *Imbalance in secretion of cholesterol and the bile salts in bile can cause cholesterol to precipitate in the gallbladder, producing **cholesterol-based gallstones,** which accounts for the most common type of **cholelithiasis.***
- *Cholelithiasis mainly arises from an **insufficiency of bile salt production,** due to several possible problems:*
  – *Hepatic dysfunction leading to decreased bile acid synthesis.*
  – *Severe ileal disease leading to malabsorption of bile salts.*
  – *Obstruction of the biliary tract.*

- *Symptoms of this condition include gastrointestinal discomfort after a fatty meal with upper right quadrant abdominal pain that persists for 1–5 hours.*
- *Probability of developing gallstones increases with age, obesity, and a high fat diet and is more prevalent in fair-skinned people of European descent, suggesting a genetic component.*

VIII. Uptake of Particles and Large Molecules by the Cell

A. Phagocytosis of large external particles, such as bacteria, occurs by **engulfment** or surrounding of the particle by the membrane.

    1. This mechanism is used mainly by specialized cells such as **macrophages, neutrophils,** and **dendritic cells.**

    2. The process starts by **binding** of the cell to the target particle.

    3. Binding is followed by **invagination** of the membrane to surround the entire particle and the membrane-encapsulated particle **pinches off** from the plasma membrane to form a **phagosome.**

    4. The phagosome then undergoes **fusion with a lysosome,** which leads to **degradation** of the engulfed material.

    5. Pinocytosis is ingestion of small particles and fluid volumes by engulfment and formation of an **endocytic vesicle.**

B. Endocytosis is a process for **uptake of specific extracellular ligands.**

    1. The process begins by **receptor-mediated binding** of target molecules or **ligands,** which are usually proteins or glycoproteins.

    2. A region of the membrane surrounding the ligand-receptor complex undergoes **invagination** by assembly of clathrin proteins on the inner face of the membrane to form a **coated pit** that encompasses the bound target.

        a. Clathrin molecules assemble into a **geometric array** that when completed forms a roughly spherical structure.

        b. The assembly forces **cooperative distortion of the membrane,** which is trapped in the interior of the clathrin coat.

    3. The structure **pinches off** the plasma membrane and forms an **endocytic vesicle,** which subsequently loses its clathrin coat.

    4. Endocytic vesicles fuse with **early endosomes,** where sorting of the endocytosed contents occurs.

        a. The acidic environment within the endosome allows **separation of receptors and their cargo (ligands).**

        b. Some **receptors are recycled** and sent back to the plasma membrane in vesicles that bud off the early endosomes.

        c. Cargo is either **targeted** for use in various areas of the cell or remains in the endosome.

        d. Remaining components form the **late endosome,** which may merge with a **lysosome,** in which the internalized materials are degraded.

    5. Examples of **receptor-mediated endocytosis** can be found in the operation of many physiologically important systems.

        a. The **transferrin receptor** is responsible for binding and internalization of iron bound to the serum protein transferrin.

        b. The availability of cell-surface receptors for hormones and growth factors is regulated through endocytosis.

        c. The **LDL receptor** binds and takes up LDL-bound cholesterol for storage or synthesis of various compounds, such as steroid hormones.

DEFECTIVE LDL RECEPTOR IN FAMILIAL HYPERCHOLESTEROLEMIA

- *Familial hypercholesterolemia (FH) results from inherited deficiency or mutation of the **LDL receptor** and consequent impairment of uptake and processing of LDL-cholesterol by the liver.*
- *LDL receptor deficiency leads to extreme hypercholesterolemia and its sequelae by two mechanisms.*
  - *Failure to take up cholesterol bound to LDL particles leads to accumulation and consequent elevation of blood LDL cholesterol.*
  - *Decreased levels of internalized cholesterol lead to elevated activity of the chief enzyme responsible for endogenous cholesterol synthesis, **HMG-CoA reductase,** and consequent excessive synthesis of cholesterol.*
- *Dramatic elevation of blood LDL-cholesterol levels in FH leads to a high risk of **atherosclerosis** at an early age due to deposition on the linings of the coronary arteries.*
- *FH is transmitted as an **autosomal dominant** trait, so even heterozygotes (frequency of 1 in 500) for LDL receptor mutations have an increased risk of atherosclerosis.*
- *The many different LDL receptor gene mutations that lead to FH can be classified into five groups according to the functional defect in the receptor:*
  - *Null alleles that produce no detectable LDL receptor protein.*
  - *Mutant receptors that become blocked during processing in the endoplasmic reticulum or Golgi apparatus and thus never reach the plasma membrane.*
  - *Mutant receptors that cannot bind LDL.*
  - *Mutant receptors that bind LDL at the cell surface but are blocked in endocytosis and thus do not internalize LDL.*
  - *LDL receptor mutants that fail to release bound LDL and do not recycle to the cell surface after internalization.*

CLINICAL PROBLEMS

A 7-year-old girl has a 1-month history of foul-smelling diarrhea. Upon further inquiry, the frequency seems to be 4–6 stools per day. She has also had trouble seeing at night in the past 2 weeks. Her WBC count is normal. Physical examination is entirely normal. Examination of a stool sample reveals that it is bulky and greasy. Analysis does not reveal any pathogenic microorganisms or parasites but confirms the presence of fats.

1. Further evaluation of this patient would likely reveal which of the following conditions?

   A. Lactose intolerance

   B. Biliary insufficiency

   C. Ileal disease

   D. Diabetes

   E. Giardiasis

A 35-year-old man is brought to the emergency department in a confused and semi-comatose state following a motor vehicle accident. His wife explains that he has type 1 diabetes mellitus. They were at a party earlier in the evening and both of them had two or three drinks. She is unsure whether he took his insulin before they left for the party. Physical examination reveals peripheral cyanosis and dehydration. While you are checking his abdomen, the patient doubles over and vomits. A fruity odor is detectable on his breath. A spot glucose reveals severe hyperglycemia.

2. Testing of the patient's urine would likely reveal abnormally high levels of which of the following?

    A. Protein

    B. Hemoglobin

    C. Acetoacetate

    D. Lactate

    E. Pyruvate

A 19-year-old man complains of "brown urine" and pain in the muscles of his arms and legs experienced while playing touch football. He has had several episodes of muscle pain during exercise, but he had not noticed darkening of his urine afterward. The pain usually resolved overnight. Physical examination reveals a well-fed male of normal stature. Reflexes and range of motion in all arms and legs are normal, but there is some paraparesis (weakness), especially in his right leg. A muscle biopsy is taken and sent for specialized testing. The patient is sent home with a recommendation to take a dietary carnitine supplement.

3. Which of the following is the most likely diagnosis?

    A. MCAD deficiency

    B. Carnitine deficiency

    C. CPT-I deficiency

    D. CPT-II deficiency

    E. Marfan syndrome

A 21-month-old girl is hospitalized with a suspected gastrointestinal virus. She is vomiting and lethargic. Physical examination reveals poor muscle tone, guarding, and some cyanosis. Blood is drawn for chemistry and complete blood count, and an intravenous line is ordered for administration of glucose and electrolytes. Before this work is completed, the patient suffers a seizure and lapses into a coma. She dies 3 days later, despite intravenous treatments to stabilize her blood sugar. The original blood sample taken on admission reveals severe hypoglycemia and hyperammonemia. An acylcarnitine profile of her blood indicates the presence of significant C6–C10 species.

4. An evaluation of this patient's liver would reveal deficiency of which of the following enzyme activities?

    A. CPT-I

    B. CPT-II

    C. Pyruvate carboxylase

    D. MCAD

    E. Pyruvate dehydrogenase

A newborn baby boy is unconscious after having suffered a seizure. A variety of dysmorphic facial features are evident, including a high forehead, a flat occiput, large fontanelles, and a high arched palate. All reflexes are depressed. There is hepatomegaly consistent with impaired liver function revealed by blood chemistry. Testing also reveals high levels of copper in the blood, but adrenal function is within normal limits. Despite all interventions, the infant dies within a week of birth. Autopsy reveals an accumulation of VLCFAs in tissue samples of the liver and kidneys.

5. Microscopic examination of tissues from this patient would likely indicate an absence of which of the following cellular components?

   A. Peroxisomes

   B. Lysosomes

   C. Mitochondria

   D. Endoplasmic reticulum

   E. Lipid droplets

A 38-year-old man with a family history of cardiovascular and cerebrovascular disease makes an appointment for a routine physical examination with a physician he has not seen before. He explains that his father died young of a heart attack and that two paternal uncles have suffered strokes in their late 40s. Physical examination reveals yellowish lumps on his eyelids (xanthelasmas, which are often associated with a lipid disorder) and a resting blood pressure of 186/95 mm Hg. There is some excess visceral fat, and his body mass index calculates to 26.5. Total serum cholesterol (476 mg/dL) and triglycerides (288 mg/dL) are elevated and subsequent angiography reveals atherosclerotic restrictions of at least two coronary arteries.

6. This patient's condition is most likely brought about by impairment of which of the following cellular functions?

   A. Synthesis of apoproteins needed for LDL assembly

   B. Production of HMG CoA reductase

   C. Vesicular trafficking mediated by the cytoskeleton

   D. Receptor-mediated endocytosis of the LDL receptor

   E. Uptake of cholesterol-derived bile salts in the intestine

ANSWERS

1. The answer is B. This patient's greasy, foul-smelling stools indicate steatorrhea. Her vision problems may be a manifestation of vitamin A deficiency due to fat malabsorption. The most likely explanation is biliary insufficiency, ie, decreased bile salt production leading to poor emulsification of dietary fats. Active ileal disease is a possibility, but the WBC count would likely be elevated unless her condition was in remission. Infection with *Giardia* is less likely due to the absence of pathogenic organisms in her stool. Lactose intolerance can produce diarrhea but not steatorrhea.

2. The answer is C. This patient appears to be suffering from diabetic ketoacidosis induced by his failure to take his insulin on schedule. Although patients with diabetes may have elevated levels of both protein and erythrocytes in urine, depending on the

degree of renal impairment, the best answer in this case is acetoacetate, a ketone body that should be highly elevated in his urine. The very low level of insulin has allowed glucagon action to run unchecked in stimulating fuel production by his adipose tissue and liver—increased gluconeogenesis, lipolysis, and ketogenesis. Urinary elevations of lactate and pyruvate are characteristic of several metabolic disorders other than diabetes.

3. The answer is D. The most likely diagnosis in this case is CPT-II deficiency, although this is apparently a fairly mild case. The patient's muscle weakness and "brown urine" (myoglobinuria) are characteristic of this disorder. CPT-I deficiency would most likely manifest as liver dysfunction. A secondary form of carnitine deficiency due to exogenous factors such as malnutrition, infection, or dialysis, is unlikely. MCAD ordinarily manifests within the first 3–5 years of life. The patient's normal stature is inconsistent with Marfan syndrome, which is characterized by tall stature and very long bones in the extremities.

4. The answer is D. This patient appears to have suffered brain damage and died of severe hypoglycemia coupled with hyperammonemia. Deficiencies of pyruvate dehydrogenase or pyruvate carboxylase would produce psychomotor retardation due to major disruption of carbohydrate metabolism. But this patient's tests reveal a key finding—the presence of medium-chain (C6–C12) fatty acylcarnitine species in her blood. This is diagnostic of MCAD deficiency, an impairment of metabolism of these fats and their accumulation to toxic levels. There has been speculation that MCAD deficiency and other undiagnosed metabolic disorders may be responsible for a significant proportion of sudden infant death syndrome (SIDS) cases. MCAD deficiency is now being tested as a component of mandatory newborn screening in many states.

5. The answer is A. This child appears to have died of a form of Zellweger syndrome. The key findings supporting this conclusion include the dysmorphic skeletal features, hepatomegaly, elevated blood copper and, most importantly, the accumulation of VL-CFAs in tissues. Zellweger syndrome is a disorder of peroxisome biogenesis, and cells of affected individuals have very small or absent peroxisomes. The other major peroxisomal disorder involving accumulation of VLCFAs, X-linked adrenoleukodystrophy, does not normally manifest during the neonatal period and is not associated with skeletal abnormalities. Further, peroxisomes are of normal size and appearance in the cells of patients with X-linked adrenoleukodystrophy.

6. The answer is D. This patient's tests indicate that he has severe hypercholesterolemia and high blood pressure in conjunction with atherosclerosis. The deaths of several of his family members due to heart disease before age 60 suggest a genetic component, ie, familial hypercholesterolemia. This disease results from mutations that reduce production or interfere with functions of the LDL receptor, which is responsible for uptake of LDL-cholesterol by liver cells. The LDL receptor binds and internalizes LDL-cholesterol, delivers it to early endosomes and then recycles back to the plasma membrane to pick up more ligand. Reduced synthesis of apoproteins needed for LDL assembly would tend to decrease LDL levels in the bloodstream, as would impairment of HMG CoA reductase levels, the rate-limiting step of cholesterol biosynthesis. Reduced uptake of bile salts will also decrease cholesterol levels in the blood.

CHAPTER 9
NITROGEN METABOLISM

I. Digestion of Dietary Proteins

A. Proteins present in foods must be **degraded** into their component **amino acids** in order to be taken up and used by the body for fuel or as building blocks for new protein synthesis.

B. Degradation of dietary proteins (**proteolysis**) is catalyzed by **proteases** in both the stomach and small intestine.

  1. Secretion of hydrochloric acid (HCl) by the gastric mucosa in response to food intake makes the stomach very **acidic.**

    a. The **low pH (~2–2.5)** promotes protein unfolding (**denaturation**), which makes them more susceptible to cleavage by proteases.

    b. Activity of **pepsin,** the main gastric protease, is optimal at this low pH.

  2. As partially digested proteins pass through the duodenum on the way to the **intestine,** they mix with **secretions** from both the **pancreas** and the **liver (bile).**

    a. These fluids, which include **bile salts** and **sodium bicarbonate** from the pancreas, neutralize the acidity to **pH >7,** which

      (1) Promotes self-cleavage of **pancreatic proteases** from their inactive zymogen forms to active enzymes.

      (2) Supports the activity of these proteases, including **trypsin, chymotrypsin,** and several aminopeptidases and carboxypeptidases.

    b. The combined actions of these enzymes digest the proteins into **free amino acids** and dipeptides.

C. Protein breakdown products are **absorbed** into intestinal epithelial cells (**enterocytes**) by various active transport processes.

  1. Once in the epithelial cells, dipeptides are further degraded to amino acids.

  2. Amino acids are then secreted into the hepatic portal circulation.

D. Removal of the amino groups from dietary amino acids allows utilization of the carbon skeletons for fuel and further use or metabolism of the amino nitrogens.

  1. In these **transamination** reactions, the amino group from the amino acid is transferred to α-ketoglutarate to form glutamate and the corresponding α-keto acid.

    a. **Pyridoxal phosphate,** the active form of vitamin B_6, is required as a coenzyme for all these reactions.

    b. The coenzyme carries the amino group during the transfer process.

  2. These steps are **reversible,** depending on the needs of the body.

3. The reactions catalyzed by two of the most important of these enzymes, alanine aminotransferase (**ALT**) and aspartate aminotransferase (**AST**) are shown below.

$$\text{Alanine} + \alpha\text{-Ketoglutarate} \rightleftarrows \text{Pyruvate} + \text{Glutamate}$$

$$\text{Aspartate} + \alpha\text{-Ketoglutarate} \rightleftarrows \text{Oxaloacetate} + \text{Glutamate}$$

    a. ALT and AST are abundant in the liver.
    b. Elevated plasma levels of ALT and AST are diagnostic of **liver disease** or **injury.**

VITAMIN B$_6$ DEFICIENCY

- *Dietary deficiency of vitamin B$_6$ leads to **impaired amino acid metabolism** in many organs, but the CNS is most severely affected.*
- *Persons with vitamin B$_6$ deficiency exhibit a spectrum of nonspecific **neurologic manifestations,** including depression, confusion, and disorientation, which may lead to convulsions in severe cases.*
- *Vitamin B$_6$ deficiency is a rare condition, but it is prevalent in persons with chronic alcoholism due to low dietary intake and impaired conversion of pyridoxine to the active coenzyme pyridoxal phosphate.*

II. Metabolism of Ammonia

    A. Processing of the amino groups of the amino acids produces **ammonia,** which is **toxic** in its free form, especially to nerve cells. So, its metabolism is designed to keep **blood levels low** (ie, <40 μM).

    B. In the liver, **glutamate dehydrogenase** catalyzes the oxidative deamination of glutamate to produce free ammonia (Figure 9–1).
        1. This reaction is reversible and utilizes different NAD-based cofactors in the forward and reverse directions.
        2. ADP and GDP, which indicate low energy levels in the cell, and ATP and GTP, which indicate high energy levels, are allosteric effectors of the enzyme (Figure 9–1).

    C. Ammonia is converted to a **nontoxic form,** mainly **glutamine,** for transport to the liver for further processing.
        1. In most tissues, **glutamine synthetase** combines free ammonia with glutamate to make glutamine (Figure 9–1).
        2. In muscle, transamination of pyruvate forms alanine, which is transported to the liver, where the reaction is reversed (the **alanine cycle**).

ACQUIRED HYPERAMMONEMIA

- *Blood levels of ammonia exceeding 40 μM have direct **neurotoxic** effects, especially disruption of neurotransmission in the CNS.*
- *Liver disease due to alcohol abuse, chronic hepatitis, or hemochromatosis, leads to impairment of ammonia disposal by the urea cycle and is often the cause of this condition in adults.*
- *Patients suffering from ammonia intoxication show major neurologic symptoms, including slurred speech; blurry vision; somnolence; lethargy; ataxic gait; tremors; vomiting; seizures and cerebral edema, which can lead to coma, brain damage, or death.*

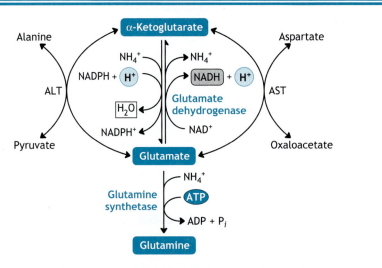

Figure 9–1. Molecular interconversions in handling of ammonia. The major enzyme responsible for interconversion of glutamate and α-ketoglutarate is glutamate dehydrogenase. No free ammonia is ever present during direct transfer of amino groups from alanine or aspartate via transamination to produce glutamate. ALT, alanine aminotransferase; AST, aspartate aminotransferase.

- *Treatment involves administration of a **low-protein diet** to minimize nitrogen burden or liver transplantation in cases where liver damage is severe.*
- *Administration of **benzoate,** phenylbutyrate, or phenylacetate can also be used to manage the condition.*
  - *Benzoate combines with glycine to form hippurate, which is excreted in the urine and decreases the overall nitrogen burden.*
  - *Phenylacetate (administered directly or by conversion from phenylbutyrate) sequesters free ammonia by combining with it to form phenylacetylglutamine, which is cleared by the kidneys.*

III. The Urea Cycle

    A. The urea cycle **converts ammonia to urea, a nontoxic substance.**

        1. One of the nitrogen atoms for urea synthesis comes from **ammonia** and the other is donated by **aspartate.**

        2. The carbon atom of urea comes from CO_2.

        3. Urea formed in the liver is highly water-soluble and is carried by the blood to the kidneys where it is **filtered and excreted in the urine.**

    B. The reactions of the urea cycle are catalyzed by five enzymes (Figure 9–2).

        1. The first two reactions occur in the mitochondria.

            a. The **rate-limiting step,** formation of carbamoyl phosphate, is catalyzed by the key enzyme, **carbamoyl phosphate synthetase I (CPS-I).**

            b. **Carbamoyl phosphate** is then **coupled to ornithine** to form **citrulline** and the cycle begins (Figure 9–2).

            c. The free ammonia that is utilized in the initial step is derived by deamination of glutamine by glutaminase or glutamate dehydrogenase.

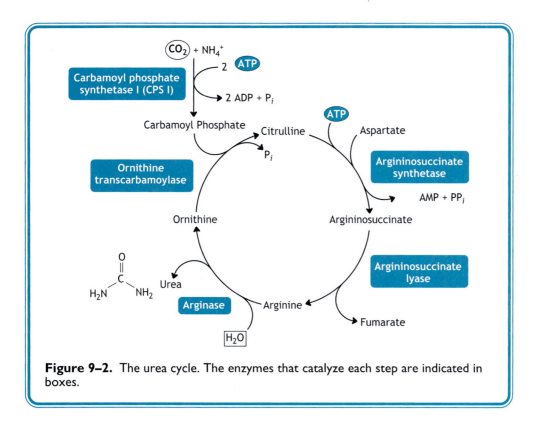

Figure 9–2. The urea cycle. The enzymes that catalyze each step are indicated in boxes.

   2. Citrulline is transported out of the mitochondria to the cytosol, where the other three reactions of the urea cycle take place.
C. Flux of ammonia through the **urea cycle is regulated** by two factors:
   1. Availability of substrates: aspartate, ammonia, and CO_2.
   2. Allosteric activation of CPS-I by *N*-acetylglutamate, which is formed from acetyl CoA and glutamate, and indicates adequate availability of substrates for the urea cycle.
D. The overall reaction of the urea cycle indicates that handling of ammonia requires expenditure of significant energy.

$$\text{Aspartate} + NH_4^+ + CO_2 + 3ATP \rightarrow \text{Urea} + \text{fumarate} + \\ 2ADP + AMP + 2P_i + PP_i + 3\,H_2O$$

HEREDITARY HYPERAMMONEMIA

* *Deficiencies of several key enzymes in the pathways for handling ammonia and synthesizing urea (Figure 9–2) are responsible for the following:*
   – *Ornithine transcarbamoylase deficiency, an X-linked condition and the most common of these disorders.*
   – *CPS-I deficiency.*
   – *Arginase deficiency, which is inherited in an autosomal recessive manner and causes a rare hyperargininemia.*

– *Argininosuccinate synthetase deficiency, which leads to citrullinemia.*
   – *Argininosuccinate lyase deficiency.*
- *Symptoms of hereditary hyperammonemia include many of the **neurologic manifestations** of acquired hyperammonemia, but they are seen mainly in **infants** and frequently lead to **mental retardation.***
- *Long-term treatment involves limiting dietary protein to minimize the ammonia burden in conjunction with strategies to decrease ammonia, eg, dialysis.*

IV. Catabolism of Amino Acids

A. The first step in metabolism of most amino acids is removal of the α-amino group by **transamination.**

B. The mechanisms of amino acid degradation are grouped according to the ways their **carbon skeletons** are subsequently metabolized.
   1. **Glucogenic** amino acids can be used for synthesis of glucose (Figure 9–3).
      a. The **glucogenic** amino acids are alanine (Ala), arginine (Arg), asparagine (Asn), aspartate (Asp), cysteine (Cys), glutamate (Glu), glutamine (Gln), glycine (Gly), histidine (His), proline (Pro), serine (Ser), methionine (Met), valine (Val), and threonine (Thr).
      b. The carbon skeletons of these amino acids are converted to **pyruvate** or one of the **tricarboxylic acid (TCA) cycle intermediates.**
      c. Depending on the energy needs of the body, these amino acids can be used **directly as fuel** or **converted to glucose** in the liver.
   2. **Ketogenic** amino acids yield acetyl CoA and acetoacetate.
      a. The ketogenic amino acids are leucine (Leu) and lysine (Lys).
      b. Lys can be converted via a complex series of nine reactions into acetoacetyl CoA; alternatively, Lys can be utilized for synthesis of **carnitine.**
   3. The carbon skeletons of isoleucine (Ile), phenylalanine (Phe), tyrosine (Tyr), and tryptophan (Trp) are both **glucogenic** and **ketogenic.**

C. The **branched-chain amino acids** Leu, Ile, and Val share a common pathway for metabolism, which occurs in the **peripheral tissues,** such as muscle, rather than in the liver (Figure 9–4).
   1. Removal of the amino groups by branched-chain amino acid transaminase forms the corresponding **α-keto acids.**
   2. The α-keto acids then undergo **oxidative decarboxylation** to their coenzyme A derivatives catalyzed by **branched-chain α-keto acid dehydrogenase.**
      a. This a multi-enzyme complex located on the inner mitochondrial membrane.
      b. The enzyme is similar in organization to the pyruvate dehydrogenase complex and utilizes thiamine pyrophosphate, lipoic acid, NAD⁺, and FAD coenzymes.
   3. At this point, the pathways for branched-chain amino acid metabolism diverge.
      a. Ile and Val are metabolized further to propionyl CoA, which yields **succinyl CoA.**
      b. Further degradation of Leu leads eventually to the **ketone body precursor, β-hydroxy-β-methylglutaryl CoA.**

MAPLE SYRUP URINE DISEASE

- *Deficiency in branched-chain α-keto acid dehydrogenase produces **high levels of the branched-chain amino acids** and their α-keto acids in the blood, causing **neurotoxic effects** and **potential brain damage.***

CLINICAL CORRELATION

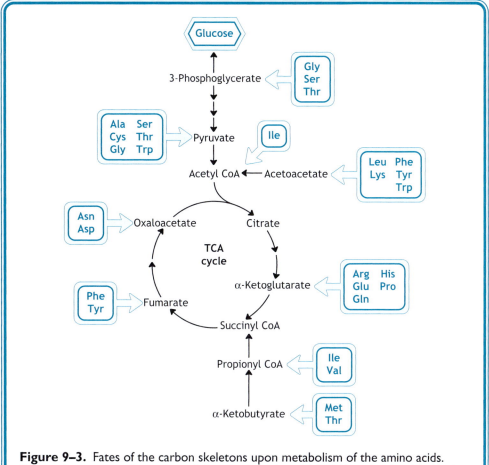

Figure 9–3. Fates of the carbon skeletons upon metabolism of the amino acids. Points of entry at various steps of the tricarboxylic acid (TCA) cycle, glycolysis and gluconeogenesis are shown for the carbons skeletons of the amino acids. Note the multiple fates of the glucogenic amino acids glycine (Gly), serine (Ser), and threonine (Thr) as well as the combined glucogenic and ketogenic amino acids phenylalanine (Phe), tryptophan (Trp), and tyrosine (Tyr). Ala, alanine; Cys, cysteine; Ile, isoleucine; Leu, leucine; Lys, lysine; Asn, asparagine; Asp, aspartate; Arg, arginine; His, histidine; Glu, glutamate; Gln, glutamine; Pro, proline; Val, valine; Met, methionine.

- The α-keto acids and their metabolic byproducts are excreted in urine, and these compounds cause the characteristic sweet, **"maple syrup" aroma.**
- Infants suffering from maple syrup urine disease exhibit failure to thrive, feeding problems, vomiting, dehydration, and severe metabolic acidosis, with **mental retardation** as a major sequela.
- Treatment of this rare, autosomal recessive disorder involves a diet low in these amino acids as well as dietary supplementation with keto acids and thiamine.

   D. Metabolism of the **aromatic amino acids**, especially Tyr and Trp, via several alternative pathways leads to synthesis of physiologically important compounds.

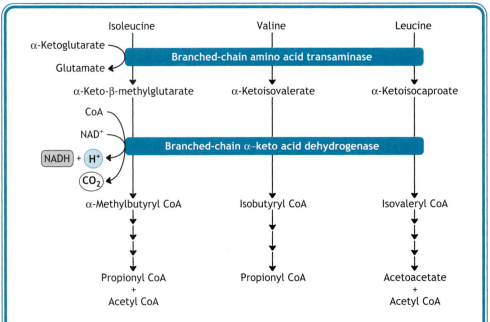

Figure 9–4. Metabolism of the branched-chain amino acids. The first two reactions, transamination and oxidative decarboxylation, are catalyzed by the same enzyme in all cases. Details are provided only for isoleucine. Further metabolism of isoleucine and valine follows a common pathway to propionyl CoA. Subsequent steps in the leucine degradative pathway diverge to yield acetoacetate. An intermediate in the pathway is 3-hydroxy-3-methylglutaryl CoA (HMG-CoA), which is a precursor for cytosolic cholesterol biosynthesis.

1. The main pathway for degradation of Tyr leads to formation of **fumarate** and **acetoacetate.**
2. Synthesis of **catecholamines** from Tyr begins with hydroxylation of the Tyr ring catalyzed by **tyrosine hydroxylase.**
   a. The primary product of this reaction is 3,4-dihydroxyphenylalanine (**DOPA**).
   b. Subsequent reactions from DOPA produce, in turn, **dopamine, norepinephrine,** and **epinephrine.**
3. Synthesis of the aromatic quinone pigment, **melanin,** is initiated by oxidation of the Tyr ring by **tyrosinase.**

ALBINISM: A DISORDER OF MELANIN PRODUCTION

• A **deficiency of tyrosinase** activity leads to a reduction in melanin production.
• The classic and most severe form of **albinism, a complete lack of melanin** in the hair and skin and of color in the iris, is due to complete deficiency of tyrosinase.
• Affected persons also show **hypersensitivity to sunburn** and **photophobia** (an aversion to sunlight) due to painful effects of light on the eyes.

4. Trp metabolism is complex and leads to multiple products, including serotonin, melatonin, and NAD^+.

a. The pathway leading from Trp to **nicotinate mononucleotide,** a precursor of NAD⁺, requires nine reactions.

    b. Hydroxylation of Trp by tryptophan 5-monooxygenase leads to production of the neurotransmitter **serotonin** (5-hydroxytryptamine), which can be converted to the sleep-inducing molecule, **melatonin.**

E. Metabolism of His begins with oxidative deamination leading to production of free ammonia and the intermediate **urocanic acid.**

    1. The enzyme responsible for this reaction is histidine ammonia lyase or **histidase.**

    2. Further metabolism leads to opening of the imidazole ring and finally production of glutamate.

    3. An alternative pathway for His metabolism is decarboxylation to produce **histamine,** the potent mediator of allergic reactions.

V. Biosynthesis of Amino Acids

A. Under normal conditions, the body has enzymes capable of synthesizing only 10 of the 20 common amino acids.

    1. The **essential amino acids** are those that cannot be made by the body and must be obtained through the diet.

    2. Arg and His are **conditionally essential;** they must be provided by the diet when the body's ability to synthesize them is outstripped, such as during periods of active growth or during recovery from illness.

    3. The **nonessential amino acids** can be synthesized by the body using carbon skeletons from metabolic intermediates or by modification of other amino acids.

B. Ala, Asp, and Glu are synthesized by transfer of an amino group to their α-keto acids.

    1. Pyruvate yields Ala.

    2. Oxaloacetate gives Asp.

    3. α-Ketoglutarate is converted to Glu.

C. Gln, Asn, and Pro share similar or overlapping pathways of synthesis in the body.

    1. Gln is made from Glu by reductive amidation (see Figure 9–1).

        a. This reaction is catalyzed by glutamine synthetase and requires ATP.

        b. This is a major mechanism for handling ammonia.

    2. A similar reaction is responsible for synthesis of Asn from Asp.

    3. Pro is derived by cyclization of Glu with subsequent reduction.

D. Ser and Gly are synthesized by modification of the glycolytic intermediate 3-phosphoglycerate.

    1. Ser is made in several steps that add an amino group and remove the phosphate.

    2. Gly is derived from Ser by removal of the carboxymethyl group of the side chain.

E. The pathway for synthesis of the sulfur-containing amino acid Cys from Met provides important compounds for other reactions (Figure 9–5).

    1. Transfer of adenosine from ATP to the sulfur of the essential amino acid Met produces **S-adenosylmethionine (SAM).**

        a. The methyl group of SAM is "activated" and carries relatively high energy.

        b. SAM is used as a **methyl donor** in many physiologic reactions.

    2. Conversion of **homocysteine** to Cys occurs in two reactions catalyzed by two pyridoxal phosphate–requiring enzymes, cystathionine β-synthase and γ-cystathionase.

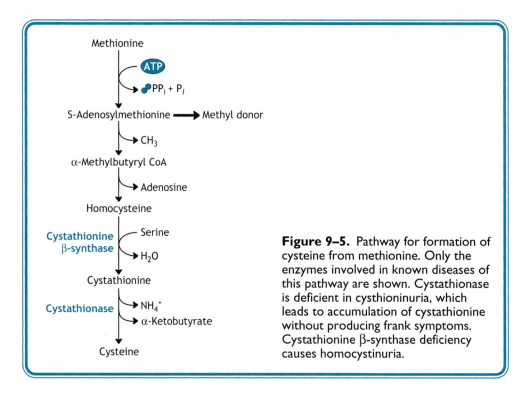

Figure 9–5. Pathway for formation of cysteine from methionine. Only the enzymes involved in known diseases of this pathway are shown. Cystathionase is deficient in cysthioninuria, which leads to accumulation of cystathionine without producing frank symptoms. Cystathionine β-synthase deficiency causes homocystinuria.

HOMOCYSTINURIA

- **Homocystinuria** is caused by inherited deficiency of one of the enzymes in the pathway from Met to Cys.
  – The major type is caused by **cystathionine β-synthase deficiency,** leading to accumulation of upstream intermediates in the pathway, especially homocysteine.
  – High levels of **homocysteine** cause direct **neurotoxic** and **teratogenic effects** in addition to the effects of Cys deficiency.
- Patients with homocystinuria have high levels of homocysteine in blood and urine, and they exhibit skeletal abnormalities such as **scoliosis, high arched palate, and generalized osteoporosis** in childhood or early adulthood.
- **Ectopia lentis** (with downward dislocation) is a characteristic finding of this disease, which may also lead to cardiovascular manifestations and mental retardation.
- Treatment for these autosomal recessive conditions is to provide a Met-restricted diet with vitamin B_6 supplementation to enhance any residual enzyme activity that may be available and with Cys supplementation to make up for the deficiency in its synthesis.

     F. Tyr is synthesized by **hydroxylation** of the phenyl ring of the essential amino acid Phe (Figure 9–6).

        1. The reaction is catalyzed by **phenylalanine hydroxylase** using molecular oxygen as the oxygen donor.

        2. Tetrahydrobiopterin donates electrons as a required **coenzyme** for the reaction.

PHENYLKETONURIA

- **Classic phenylketonuria (PKU)** is due to **deficiency of phenylalanine hydroxylase,** which leads to brain damage due to the **neurotoxic effects** of accumulated Phe or its metabolites.

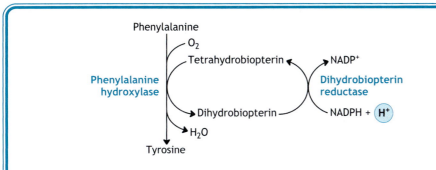

Figure 9–6. Synthesis of tyrosine from phenylalanine. Hydroxylation of phenylalanine to tyrosine is one of several reactions in the body that require tetrahydrobiopterin as a cofactor to provide electrons and hydrogen as reducing equivalents.

- *Infants with PKU have* **hyperphenylalaninemia** *with spillover of metabolic products into the urine producing a musty odor.*
- *This is an autosomal recessive disorder that is the most common inborn error of amino acid metabolism, with an incidence of 1 in 11,000 live births in the United States.*
- *If not treated within the first week of life, PKU causes* **mental retardation,** *developmental delays, and microcephaly.*
- *Mandatory newborn screening for PKU has allowed early detection and mitigation of the most severe effects in most cases.*
- *When detected within a few days of birth, a* **diet low in Phe** *is established and should be maintained until at least 8 years of age.*
  - *This helps prevent neurologic effects during development of the nervous system.*
  - *Recent evidence suggests possible progressive deterioration in mental function, eg, declining IQ after adult PKU patients suspend the Phe-restricted diet.*
  - *Dietary supplementation with Tyr is necessary to make up for decreased synthesis.*
- *Mild or* **nonclassic forms of PKU** *can be caused by deficiency of dihydrobiopterin reductase.*
- ***Maternal PKU*** *occurs when a pregnant woman with uncontrolled PKU has high levels of Phe in her blood, leading to elevated levels of Phe in fetal blood and consequent neurologic damage, including microcephaly and mental retardation.*
  - *Pregnant women with PKU should maintain a low-Phe diet to avoid inducing neurologic damage and potential mental retardation in the fetus, since fetal phenylalanine hydroxylase activity acquired from the father would be inadequate to metabolize the mother's high plasma Phe.*

VI. Porphyrin Metabolism

   A. Porphyrins are nitrogen-containing, cyclic compounds that bind metal ions in coordination complexes, ie, **metalloporphyrins.**
   1. The main metalloporphyrin in the body is **heme,** which has a ferrous Fe^{2+} iron atom coordinated by **protoporphyrin IX.**
   2. Heme synthesis takes place in all cells, but most of the body's heme is made in the liver and bone marrow.
      a. Fully 85% of the body's heme is synthesized by the **erythropoietic cells** of the bone marrow for **hemoglobin** production.
      b. The **liver** synthesizes more heme than most other organs in order to maintain its high content of **cytochromes.**

B. Heme is synthesized from **glycine** and **succinyl CoA precursors** via a complex series of reactions (Figure 9–7).

  1. Condensation of Gly and succinyl CoA to produce **δ-aminolevulinic acid (ALA)** is the **rate-limiting step** of heme biosynthesis.

  a. The reaction is catalyzed by **ALA synthase,** which requires pyridoxal phosphate as a coenzyme.

  b. This reaction and the last three reactions in the series occur in the mitochondria, while those in between take place in the cytoplasm.

  c. Two molecules of ALA combine in the next step to produce **porphobilinogen.**

  d. Four molecules of porphobilinogen are condensed together with subsequent modification to produce protoporphyrin IX, which takes on Fe^{2+} to form **heme.**

  2. When the amount of heme available exceeds the amount of the core proteins, ALA synthase in the liver is inhibited to reduce flux through the pathway.

  a. In this case, the ferrous iron in heme is oxidized to the **ferric (Fe^{3+}) state** forming **hemin.**

  b. It is **hemin** that inhibits ALA synthase both by feedback using an **allosteric mechanism** and also by regulating **ALA synthase gene expression** by repressing its transcription.

  c. In contrast to the effect of hemin, many drugs, including **barbiturates,** stimulate expression of the ALA synthase gene in the liver, leading to upregulation of enzyme activity.

  3. Regulation of heme synthesis in erythroid cells occurs at enzymes catalyzing several steps of the pathway, including ferrochelatase.

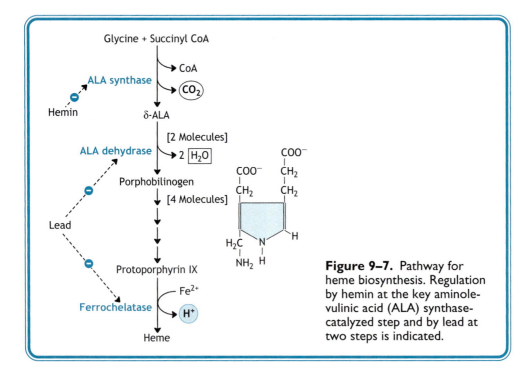

Figure 9–7. Pathway for heme biosynthesis. Regulation by hemin at the key aminolevulinic acid (ALA) synthase-catalyzed step and by lead at two steps is indicated.

LEAD POISONING INHIBITS HEME BIOSYNTHESIS

- *Lead poisoning produces a **microcytic anemia** that arises from the ability of lead to block erythropoiesis by **inhibiting heme synthesis** in the bone marrow at two steps.*
  - *– **Lead inhibits ALA dehydrase,** which blocks the condensation of ALA molecules into porphobilinogen.*
  - *– **Lead ions also inhibit ferrochelatase**-catalyzed insertion of Fe^{2+} into protoporphyrin IX in the final step of heme synthesis.*
- *Symptoms of lead toxicity include **gastrointestinal effects,** such as vomiting, constipation, abdominal pain, and appetite loss, as well as **neurologic effects,** which manifest in children as decreased attention span, behavioral problems, and apparent learning disorders.*
- *Patients may also exhibit pallor and suffer from weakness and lethargy due to anemia.*

PORPHYRIAS: INHERITED DEFECTS IN HEME SYNTHESIS

- *The porphyrias are a heterogeneous group of **diseases of porphyrin metabolism** characterized by a variety of dermatologic, neurologic, and psychological manifestations.*
- *Common features include **photosensitivity** of the skin, which causes itching and burning sensations (pruritus), rashes, and blisters.*
- ***Chronic porphyria,** or porphyria cutanea tarda, is the most common form of the disease.*
  - *– This condition produces photosensitivity, episodic skin symptoms, and pink-tinted urine, which darkens upon standing in air due to excretion of porphobilinogen, an intermediate in heme synthesis.*
  - *– These effects are due to **deficiency of uroporphyrin decarboxylase,** one of the enzymes of the heme synthetic pathway.*
- *Several types of acute liver porphyrias produce episodic attacks of neurologic and psychological symptoms as well as gastrointestinal distress.*
  - *– This group includes acute intermittent porphyria, hereditary coproporphyria, and variegate porphyria.*
  - *– These diseases arise from excessive expression of ALA synthase, coupled with deficiencies in other enzymes that operate downstream in the heme synthesis pathway.*
  - *– In many of these conditions, inhibition of heme formation leads to up-regulation of ALA synthase and exacerbates the accumulation of toxic intermediates.*
- *Treatment of these disorders is symptomatic and supportive during acute episodes.*
  - *– Patients should avoid sunlight and drugs such as barbiturates and ethanol, which induce cytochromes in the liver, further reducing available heme and up-regulating ALA synthase expression.*
  - *– Severe cases can be managed to some extent by administration of hemin, with the goal of suppressing ALA synthase expression.*

C. Turnover of RBCs by the **reticuloendothelial (RE) system** (peripheral macrophages, spleen, and liver) and degradation of cytochromes in cells throughout the body necessitates the **degradation of heme.**

1. Within cells of the RE system, heme is degraded to bilirubin in a two-step process (Figure 9–8).
   a. The porphyrin ring is opened and the iron atom is removed by the action of heme oxygenase to produce the green-colored intermediate **biliverdin.**
   b. Subsequent reduction converts biliverdin to **bilirubin,** which has a red-orange color.
   c. Bilirubin is not very water-soluble, so most of it is carried to the liver bound to **albumin.**

2. In cells of the liver, bilirubin undergoes modification to increase its water solubility so that it can be excreted more easily.
   a. Bilirubin is conjugated to two molecules of glucuronic acid, creating **bilirubin diglucuronide.**

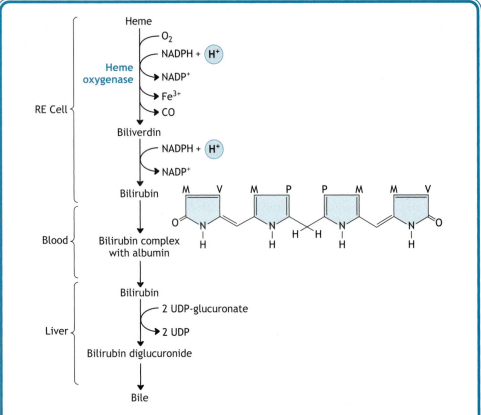

Figure 9–8. Pathway for metabolism of heme and excretion as bilirubin. Heme degradation begins with heme oxygenase, which catalyzes a complex set of reactions that simultaneously open the protoporphyrin ring structure and release iron in the ferric (Fe^{3+}) state. This is the only physiologic reaction that makes endogenous CO in the body; a portion of the small amounts made is expired via the lungs. The structure of the main form of bilirubin is shown. Symbols for the side groups indicate M, methyl; V, vinyl; P, propyl. Formation of the diglucuronide is catalyzed by bilirubin uridine diphosphate (UDP) glucuronyltransferase. RE, reticuloendothelial.

     b. Bilirubin diglucuronide is transported out of the hepatocytes into the bile canaliculi and is thus **excreted in bile.**

  3. Within the intestine, bacteria convert bilirubin to **urobilinogen.**

     a. Most of the urobilinogen is oxidized to **stercobilin,** which turns the **feces brown.**

     b. Some urobilinogen is reabsorbed and transported through the bloodstream to the kidneys, where it is excreted as **urobilin,** which makes **urine yellow.**

JAUNDICE

• *Jaundice (or icterus) is defined as **yellowing** of the skin, sclerae and fingernail beds due to increased concentration of bilirubin in the blood (**hyperbilirubinemia**).*

CLINICAL
CORRELATION

- *Jaundice is not itself a disease but is an important diagnostic indicator of many underlying conditions.*
- *In newborns, jaundice can lead to toxic encephalopathy due to deposition of bilirubin within the lipid regions of membranes of the brain (**kernicterus**).*
- *The **etiology of bilirubinemia** can be loosely classified into three types: hemolytic, obstructive, and hepatocellular.*
- ***Hemolytic jaundice** arises as a consequence of **excessive** destruction of RBCs.*
  - *This overloads the capacity of the RE system to metabolize heme.*
  - *Failure to conjugate bilirubin to glucuronic acid causes accumulation of bilirubin in the unconjugated form in the blood.*
- ***Obstructive jaundice,** as the name implies, is caused by blockage of the bile duct by a gallstone or a tumor (usually of the head of the pancreas).*
  - *This prevents passage of bile into the intestine and consequently conjugated bilirubin builds up in the blood.*
  - *Patients with this condition suffer **severe abdominal pain** associated with the obstruction (if due to a gallstone) and their feces are gray in color due to lack of stercobilin.*
- ***Hepatocellular jaundice** arises from liver disease, either inherited or acquired.*
  - *Liver dysfunction impairs conjugation of bilirubin.*
  - *Consequently, unconjugated bilirubin spills over into the blood.*
  - *In addition, urobilinogen is elevated in the urine.*

CLINICAL PROBLEMS

A 44-year-old woman is brought to the emergency department doubled over with abdominal pain. Her husband states that the pain began several hours earlier, comes in waves, but has not really subsided completely even for brief periods since then. Oral antacids have not helped the pain at all. Her discomfort is not relieved by defecation. A stool sample is light gray or clay-colored. Physical examination shows right upper quadrant abdominal pain. Her sclerae are slightly yellow in color. A sonogram shows a 2-cm mass in the region of the bile duct.

1. Testing of her serum would be expected to reveal elevated levels of which of the following?
    A. Albumin-bound bilirubin
    B. Porphobilinogen
    C. Free, unconjugated bilirubin
    D. Conjugated bilirubin
    E. Biliverdin

A 3-day-old boy is brought to the emergency department 1 day after postnatal discharge from the hospital. His mother is concerned that he is lethargic and irritable. He has been crying nearly non-stop for about 8 hours and is inconsolable. The baby has not eaten in 16 hours, since throwing up his last feeding. Physical examination indicates poor muscle tone and grunting breath sounds. Family history reveals that his older brother died of an unknown cause after lapsing into a coma. Follow-up testing included a check of his serum ammonia level, which was 120 μM.

2. This infant is most likely suffering from a defect in which of the following processes?

    A. Pentose phosphate pathway

    B. Urea cycle

    C. Heme synthesis

    D. Amino acid uptake

    E. Cysteine synthesis

A 7-day-old girl has had a seizure. The mother explains that the baby has been vomiting and having difficulty feeding for the past 2 days. There is also a strange, sweet smell to her diapers. Physical examination is unremarkable, except for indications of dehydration. Serum test results show normal levels of glucose and ammonia. Urinalysis reveals the presence of α-keto-isocaproate and α-keto-isovalerate.

3. Evaluation of this patient's liver would reveal deficiency of which of the following enzymes?

    A. Alanine aminotransferase (ALT)

    B. Glutamine synthetase

    C. Cystathionine β-synthase

    D. Carbamoyl phosphate synthetase

    E. α-Keto acid dehydrogenase

An 11-year-old girl is being evaluated for a possible learning disorder. Recent testing indicates her IQ is 73. She reports having problems seeing in class and frequently finds it necessary to squint as she reads. Physical examination reveals lens dislocation (downward orientation); long, thin fingers and toes (arachnodactyly); high arched palate; and mild scoliosis. A series of radiographs indicates generalized osteoporosis.

4. An evaluation of this patient's blood and urine would most likely reveal an elevated level of which of the following compounds?

    A. Methionine

    B. Cysteine

    C. Lactate

    D. Cystathionine

    E. Urea

A 5-year-old boy from a family that relocated to the area 6 months ago is evaluated for symptoms of fatigue and nausea culminating in several bouts of vomiting over the past day. Physical examination indicates a slight boy with pallor and weakness in the extremities. Complete blood count and hematology reveal microcytic anemia. Further inquiries elicit the information that the family now lives in a house that was built in 1955 and they have been doing home improvement projects, including repainting and refinishing the woodwork.

5. This patient's anemia is likely due to inhibition of which of the following enzymes of heme metabolism?

    A. ALA synthase

    B. Porphobilinogen synthase

C. ALA dehydrase

D. Heme oxygenase

E. Protoporphyrinogen oxidase

A 37-year-old woman arrives in the emergency department in a confused and highly agitated state, complaining of severe abdominal pain that usually "comes and goes," but has persisted over the past few days. Her husband explains that she is normally a very level-headed, organized person, but over the past week, she has "done some strange things and is a bit irrational." Upon more detailed questioning, he admits that she has had milder episodes of this kind periodically since they have been married. Physical examination reveals a woman of medium build. Blood pressure is 156/88 mm Hg and pulse is 80 bpm. Her arms and legs are weak with reduced reflexes. Abdominal examination reveals some diffuse tenderness, but no masses are detected. Normal bowel sounds are heard. Radiographs, ultrasonograms, and endoscopic findings do not reveal any evidence of a gastrointestinal tumor or stones. Urinalysis indicates a slight pink color in the fresh sample, which became medium brown by the time it was received in the laboratory.

6. Chemical analysis of this patient's urine would likely reveal the presence of which of the following?

A. δ-Aminolevulinate

B. Bilirubin diglucuronide

C. Protoporphyrin IX

D. Porphobilinogen

E. Hemin

ANSWERS

1. The answer is D. Results of the sonogram indicate the presence of a mass that may be obstructing this patient's bile duct, which would account for her pain and the jaundice. In obstructive jaundice, bilirubin transport to the liver and its conjugation are normal, but conjugated bilirubin cannot pass into the intestine and backs up. This leads to elevation of conjugated bilirubin, the diglucuronide form, in blood. The gray color of her stool confirms the absence of the end product of bilirubin metabolism by the intestinal flora, stercobilin. Albumin-bound and free, unconjugated bilirubin are the forms bilirubin takes on its way to the liver from peripheral RE tissues, but the processes for handling these types of bilirubin are not affected in obstructive jaundice. Elevations in these types of bilirubin might be expected in hemolytic or hepatocellular jaundice, which either overwhelm or impair hepatic bilirubin conjugation. Porphobilinogen and biliverdin are intracellular intermediates in heme synthesis and degradation, respectively, that are not associated with jaundice.

2. The answer is B. All the findings are consistent with a diagnosis of hyperammonemia. A clue to its hereditary etiology is provided by the family history suggesting that the patient's sibling may have died of a similar condition. It is likely that the patient is suffering from a deficiency of one of the enzymes of the urea cycle; the most common

condition is ornithine transcarbamoylase deficiency. Although enzyme deficiencies in some of the other listed pathways may account for some of this patient's symptoms, including irritability, lack of appetite, lethargy and vomiting, none can explain the blood ammonia level, which is elevated to a range known to produce these toxic effects.

3. The answer is E. This infant appears to be suffering from maple syrup urine disease. This is consistent with all the symptoms and is strongly supported both by the subjective, "sweet" smell of the baby's urine and the finding of α-keto acids in her urine. This condition is produced by inherited deficiency of α-keto acid dehydrogenase, a key enzyme in degradation of the branched-chain amino acids. None of the other enzyme deficiencies listed can account for this unique set of symptoms. Some cases of alanine aminotransferase (ALT) deficiency have been reported but only in association with viral infections and there were no symptoms attributable to reduced ALT activity. Glutamine synthetase deficiency is a rare, autosomal recessive disorder that has produced severe malformations of the brain, multiple organ failure, and neonatal death in confirmed cases. Cystathionine β-synthase deficiency is key to homocystinuria. Deficiency of carbamoyl phosphate synthetase produces hyperammonemia, which is not observed in this case.

4. The answer is A. The constellation of symptoms exhibited by this patient is characteristic of homocystinuria. The impairment of her cognitive function could be attributed to many conditions, but the key findings are ectopia lentis with downward lens dislocation and osteoporosis in a female of this age. Homocystinuria is produced by inherited deficiency of one of the enzymes in the pathway of Met conversion to Cys. The most common form is cystathionine β-synthase deficiency, which results in accumulation of all upstream components of the pathway, including homocysteine, which is responsible for the toxic effects, and Met, which becomes elevated in the blood. Cystathionine and cysteine, which are both downstream of the block in the pathway caused by cystathionine β-synthase deficiency, would be decreased. Metabolic pathways for lactate and urea are not involved in this disease mechanism.

5. The answer is C. The patient's symptoms represent a composite of neurologic and gastrointestinal dysfunction, which are consistent with the anemia that is due to lead poisoning. Testing for lead would be appropriate for the patient, the other members of the household, and the house itself. Inorganic lead produces the microcytic anemia by inhibition of heme synthesis in erythropoietic cells of the bone marrow. All the other options represent enzymes of heme synthesis or degradation, but none of them are affected by lead.

6. The answer is D. This patient exhibits a range of symptoms that could be attributed to many different causes, but the key finding is the pink-to-brown coloration of her urine with confirmation by chemistry tests indicating the presence of porphobilinogen. This should lead to a preliminary diagnosis of one of the porphyrias. Follow-up testing would be needed to determine which of this diverse group of diseases is responsible. Initial findings and symptoms are consistent with acute intermittent porphyria, which is frequently caused by deficiency of porphobilinogen deaminase, causing porphobilinogen to accumulate and spill over into the urine. Protoporphyrin IX and hemin are downstream products of the heme biosynthesis pathway, and if anything, would be reduced in availability in this patient. The bilirubin metabolism in this patient should be normal and bilirubin conjugates are not normally present in urine. The second best guess is ALA, which may be somewhat elevated in tissues due to the block at the deaminase step, but this would not spill over into the urine.

CHAPTER 10
NUCLEIC ACID METABOLISM

I. Structures and Functions of Nucleotides

A. A **nucleoside** is formed from the linkage of a **sugar** with a nitrogen-containing **base.**
  1. The bases that make up the physiologically relevant nucleosides all have ring structures.
      a. The **purines** adenine, guanine, and inosine have a double-ring system.
      b. The **pyrimidines** cytosine, thymine, and uracil have six-membered ring structures.
  2. **Ribose and 2-deoxyribose** are the main **sugars** found in nucleosides and nucleotides.
B. **Nucleotides** are nucleosides to which one, two, or three **phosphate groups** have been added to the sugar.
  1. The bonds that interconnect the phosphate groups yield **high energy** when they are broken or hydrolyzed.
      a. ATP is the main form for provision of energy to make these bonds.
      b. Hydrolysis of these bonds is often used to drive **biosynthetic reactions.**
  2. Many **coenzymes** are **AMP** or **ADP derivatives,** including NAD$^+$, NADP$^+$, FAD, and coenzyme A.
  3. **ATP** and sometimes **GTP** can serve as a **phosphate donor** in many physiologic reactions because they have a high **phosphoryl transfer potential.**
C. Nucleotides have many important functions in the body.
  1. Nucleotides serve as the **building blocks** for synthesis of the nucleic acids **DNA** and **RNA.**
  2. Nucleotides are components of several **coenzyme** structures and of ATP.
  3. Nucleotides control rates of many enzyme-catalyzed reactions by feedback and **allosteric regulation.**
  4. In cyclic forms, such as cyclic AMP and cyclic GMP, nucleotides serve critical roles in cellular **signaling mechanisms.**

II. Biosynthesis of Purines

A. The cell maintains an important pool of purine nucleotides for synthesis of coenzymes and precursors for DNA and RNA and to support reactions that are coupled to ATP hydrolysis.
B. Purine nucleotides can be synthesized de novo from **amphibolic or dual-purpose intermediates,** which may be derived either from anabolic or catabolic pathways.

1. **Ribose 5-phosphate** derived from the pentose phosphate pathway or from dietary sources is the **starting material** that eventually gives rise to **inosine monophosphate (IMP)** (Figure 10–1).
2. The overall strategy is to build the carbon-nitrogen skeleton of a purine ring system in a **12-step process** directly on the sugar-phosphate starting material.
   a. The first step creates the multi-purpose intermediate **5′-phosphoribosyl-1-pyrophosphate (PRPP).**

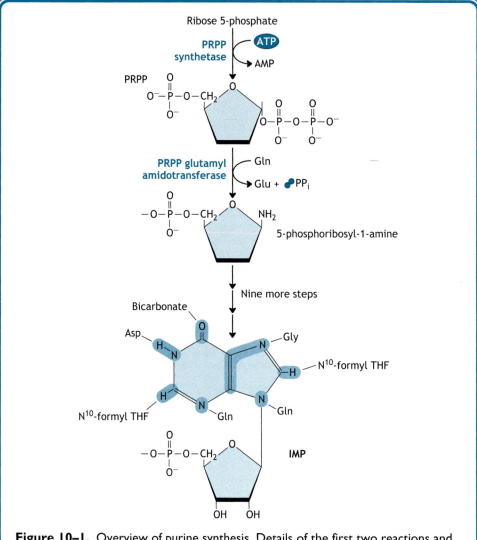

Figure 10–1. Overview of purine synthesis. Details of the first two reactions and sources of the atoms of the purine ring in inosine 5′-monophosphate (IMP) are shown. PRPP, 5′-phosphoribosyl-1-pyrophosphate; Gln, glutamine; Gly, glycine; Asp, aspartate; THF, tetrahydrofolate.

b. Then, **PRPP glutamyl amidotransferase,** the key regulatory enzyme, acts upon PRPP to begin making the purine ring; this is the **committed step** of purine synthesis.

c. **Carbons** are added to the growing ring in several ways:
   (1) By one-carbon transfer by enzymes that use **tetrahydrofolate (THF)** coenzymes.
   (2) By **incorporation of glycine** in the structure.
   (3) By addition of CO_2 in the form of bicarbonate.

d. **Nitrogens** are added by the following:
   (1) Aminotransfer reactions with **glutamine** as donor.
   (2) In a two-step mechanism with **aspartic acid** as donor.

e. Conversion of the main product of de novo synthesis, **IMP, to GMP or AMP,** occurs in two reactions, both of which are inhibited by feedback regulation by the end products.

f. Overall flux through the purine synthetic pathways is regulated primarily by **feedback inhibition** of PRPP glutamyl amidotransferase, by IMP, AMP, and GMP (Figure 10–2).

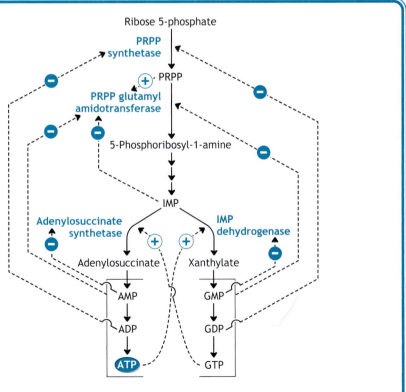

Figure 10–2. Regulation of purine synthesis by the nucleotides and the intermediate, 5′-phosphoribosyl-1-pyrophosphate (PRPP). Both feedback and feed-forward mechanisms are utilized in this intricate scheme. IMP, inosine monophosphate.

FOLIC ACID DEFICIENCY

CLINICAL CORRELATION

- *Decreased levels of folate coenzymes needed for various reactions of de novo purine synthesis and thymine synthesis produce **shortages of deoxyribonucleotides** and consequent **impaired DNA synthesis** in many tissues.*
- *Blood levels of folic acid may become inadequate due to dietary insufficiency or poor absorption due to intestinal problems or alcoholism.*
- *Folate coenzyme concentrations may also decline as a result of treatment with drugs that inhibit dihydrofolate reductase, eg, methotrexate.*
- *Patients with folic acid deficiency may have diarrhea and nausea, but the principal symptoms are weakness and easy fatigability due to **megaloblastic anemia** arising from impaired cell division in the bone marrow.*
- *Folate deficiency during pregnancy is a major contributor to **neural tube defects** because of the critical role of folate in neuronal development.*
- *Folate supplementation of food in the United States is expected to reduce folate-associated birth defects by up to 70%.*

    C. Formation of **deoxyribonucleotides** by reduction of the 2′-hydroxyl group of the ribose sugars on the ribonucleoside diphosphates ADP and GDP is catalyzed by **ribonucleotide reductase** (Figure 10–3).
      1. Thioredoxin serves as an electron donor for the reduction.
      2. The enzyme is also responsible for converting cytidine diphosphate (CDP) to 2′-dCDP and uridine diphosphate (UDP) to 2′-dUDP for use in making nucleotides for DNA synthesis.
      3. Regulation of ribonucleotide reductase by both positive feedback from ATP and negative feedback by various 2′-deoxynucleoside triphosphates (eg, dATP) is tightly coupled to the **need for DNA synthesis.**

III. Biosynthesis of Pyrimidines

    A. The **pyrimidine ring is synthesized** first and is **then attached to ribose 5-phosphate** to eventually produce the nucleotide uridine 5′-monophosphate (UMP).
      1. The first step in the pathway is synthesis of **carbamoyl phosphate** (Figure 10–4).

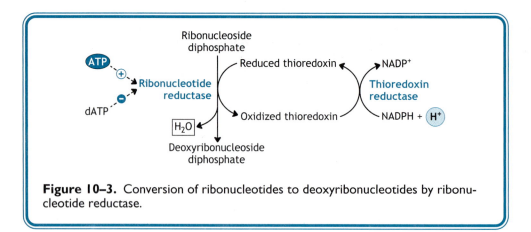

Figure 10–3. Conversion of ribonucleotides to deoxyribonucleotides by ribonucleotide reductase.

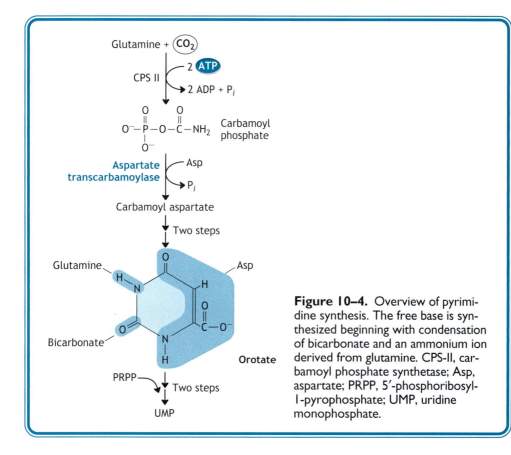

Figure 10–4. Overview of pyrimidine synthesis. The free base is synthesized beginning with condensation of bicarbonate and an ammonium ion derived from glutamine. CPS-II, carbamoyl phosphate synthetase; Asp, aspartate; PRPP, 5′-phosphoribosyl-1-pyrophosphate; UMP, uridine monophosphate.

    a. An **ammonium ion contributed by glutamine** is combined with **bicarbonate** (derived from dissolved CO_2) in a two-step reaction that requires hydrolysis of two molecules of ATP.

    b. This complex reaction is catalyzed by **carbamoyl phosphate synthetase II (CPS-II).**

    c. **CPS-II, the critical enzyme** regulating the pyrimidine synthetic pathway, is activated by ATP and PRPP and feedback-inhibited by the end product UTP.

    d. **CPS-II, a cytosolic enzyme,** is different from the mitochondrial enzyme of the urea cycle CPS-I.

2. Condensation of carbamoyl phosphate with aspartate brings together all the atoms needed to make the main pyrimidine ring (Figure 10–4).

3. Closure of the ring followed by a reduction step leads to formation of the pyrimidine base **orotate.**

4. Orotate is then connected to ribose 5-phosphate and decarboxylated to produce **UMP.**

OROTIC ACIDURIA

- *Mutation of one of the two enzyme activities of **UMP synthase** leads to orotic aciduria, characterized by accumulation of its first substrate orotic acid and insufficient levels of the product UMP, which reduces availability of uridine triphosphate (UTP) and cytidine triphosphate (CTP) for use in nucleic acid synthesis.*
- *Patients with orotic aciduria excrete large amounts of orotic acid in their urine, and they exhibit lethargy, weakness, severe anemia, and growth retardation.*
- *This autosomal recessive disorder can be treated by feeding a **diet rich in uridine,** which is salvaged to UMP and finally to UTP.*

    B. Synthesis of **UTP** and **CTP** occurs via **phosphorylation** of UMP and **interconversion** of the bases.
        1. UTP is formed from UMP by two reactions catalyzed by **nucleotide kinases** that use ATP as the phosphate donor.

$$UMP + ATP \rightarrow UDP + ADP$$
$$UDP + ATP \rightarrow UTP + ADP$$

        2. CTP is formed by modification of UTP in two steps by replacement of the carbonyl group with an amino group donated by glutamine.
    C. **Deoxythymidylate (dTMP)** is formed from 2′-deoxyuridylate (dUMP) in a one-carbon transfer by **thymidylate synthetase** (Figure 10–5).
        1. The donor coenzyme for the one-carbon transfer is N^5,N^{10}-**methylene tetrahydrofolate** (N^5,N^{10}-methylene THF); simultaneous reduction to a methyl group leaves dihydrofolate (DHF) as byproduct.
        2. N^5, N^{10}-methylene THF is regenerated from DHF by a series of reactions, one of which involves **dihydrofolate reductase.**

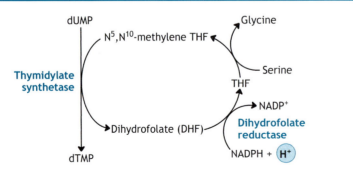

Figure 10–5. Conversion of deoxyuridylate (dUMP) to deoxythymidylate (dTMP) by thymidylate synthetase. The importance of folate coenzymes in this reaction is illustrated. NADPH + H⁺ provide the necessary reducing equivalents and serine is the source of one-carbon units present on N^5,N^{10}-methylene tetrahydrofolate (THF).

INHIBITORS OF dTMP SYNTHESIS AS ANTICANCER AGENTS

- *Several drugs that interfere with production of dTMP by blocking the reaction catalyzed by thymidylate synthetase are inhibitors of DNA synthesis and cell proliferation.*
- ***Methotrexate*** *is a folate analog that acts as a potent competitive inhibitor of dihydrofolate reductase, causing a decreased supply of THF coenzymes needed by thymidylate synthetase.*
- *The thymine analog **5-fluorouracil** (5-FU) is converted to 5-fluoro-dUMP, which acts as a suicide inhibitor of thymidylate synthetase.*

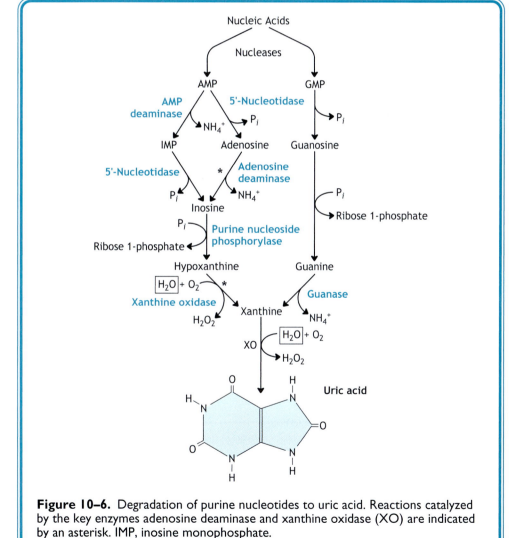

Figure 10–6. Degradation of purine nucleotides to uric acid. Reactions catalyzed by the key enzymes adenosine deaminase and xanthine oxidase (XO) are indicated by an asterisk. IMP, inosine monophosphate.

IV. Degradation of Purine and Pyrimidine Nucleotides

A. Purine nucleotides are degraded or disassembled to **uric acid** (Figure 10–6).

   1. **5′-Nucleotidase** dephosphorylates several types of ribonucleotides to form the corresponding nucleosides.
   2. AMP is deaminated by AMP deaminase to form IMP.
   3. Adenine is deaminated to inosine by the action of **adenosine deaminase.**

SEVERE COMBINED IMMUNODEFICIENCY

- Severe combined immunodeficiency (SCID) is characterized by **impaired B and T lymphocyte function** that makes affected persons vulnerable to life-threatening infections.
- One form of SCID is caused by inherited **insufficiency of adenosine deaminase** leading to accumulation of its substrate, deoxyadenosine, which is derived from DNA degradation and is converted to dATP, an allosteric inhibitor of ribonucleotide reductase.
- **Inhibition of ribonucleotide reductase** prevents synthesis of deoxyribonucleotides and thereby **blocks DNA synthesis,** an effect to which lymphoid tissues are especially susceptible.
- Treatment modalities for this autosomal recessive condition include transfusion of whole blood, bone marrow transplantation, enzyme replacement, and gene therapy.

   4. Phosphorylases remove the ribose sugars to yield the bases guanine or hypoxanthine (from adenine or inosine nucleosides).
   5. **Xanthine oxidase** catalyzes two consecutive reactions that result in formation of uric acid (Figure 10–6).
      a. Both reactions involve oxidation of ring carbons on the base.
      b. Uric acid is ultimately excreted in urine.

GOUT

- **Hyperuricemia** and chronic or episodic joint pain due to deposition of sodium urate crystals and consequent **inflammation (gouty arthritis)** are the hallmarks of gout.
- Uric acid is minimally water-soluble and most cases of gout arise from inadequate excretion by the kidneys (so-called "underexcretors"), leading to build-up of uric acid and precipitation of **urate stones** in the kidneys and extremities.
- The joints of the hands and feet are prone to accumulation of crystals because of reduced solubility of sodium urate at the slightly cooler temperature of the extremities.
- **Primary gout** can be caused by overproduction of purine catabolites due to X-linked mutations of PRPP synthetases that render the enzyme insensitive to allosteric inhibitors.
- Excessive eating and consumption of food with high levels of purine nucleotides, such as meats and grains, exacerbates gout by elevating uric acid levels.
- Treatment of gout depends on the severity of the patient's condition.
  - Acute attacks with extensive build-up of uric acid as soft, nodular masses in the joints (**tophaceous gout**) are treated with **colchicine** to reduce inflammation.
  - **Uricosuric agents, such as probenecid,** enhance uric acid excretion via the kidneys.
  - Treatment with **allopurinol,** an inhibitor of xanthine oxidase, leads to decreased uric acid production.
  - Elevated levels of xanthine and hypoxanthine caused by allopurinol treatment are tolerated because they remain soluble and are excreted by the kidneys.

B. Pyrimidine nucleotides are degraded progressively to **β-amino acids,** ie, amino acids that have the amino group on the β rather than α carbon.

1. The nucleotides are initially converted to nucleosides by nonspecific **phosphatases** that remove the phosphate groups.
2. Cytidine is **deaminated** to uridine and deoxycytidine is deaminated to deoxyuridine.
3. The sugars are removed from the nucleosides to form the **bases uracil** and **thymine.**
4. These **bases are then degraded** to two β-amino acids, uracil to β-**alanine** and thymine to β-**aminoisobutyrate.**
   a. The same enzymes catalyze these parallel reactions on either substrate.
   b. β-Aminoisobutyrate is derived exclusively from DNA, so the amount excreted in urine is a measure of DNA turnover.
   c. β-Alanine and β-aminoisobutyrate are converted into malonyl-CoA and methylmalonyl-CoA for further metabolism.

V. Salvage Pathways

A. **Salvage pathways** allow synthesis of nucleotides from free purines or pyrimidines that arise from nucleic acid degradation or dietary sources, which is more **economical for the cell** than de novo synthesis.

B. Free purines may be joined with PRPP to produce mononucleotides by one of two enzymes.
   1. The reaction shown below is catalyzed by adenine phosphoribosyltransferase **(APRT).**

$$Adenine + PRPP \rightarrow AMP + PP_i$$

   2. The following reactions are catalyzed by hypoxanthine-guanine phosphoribosyltransferase **(HGPRT).**

$$Hypoxanthine + PRPP \rightarrow IMP + PP_i$$
$$Guanine + PRPP \rightarrow GMP + PP_i$$

C. Salvage of purine nucleosides is achieved by phosphorylation with ATP as the phosphate donor.

LESCH-NYHAN SYNDROME

- *Lesch-Nyhan syndrome is an **X-linked** disorder arising from **deficiency of HGPRT,** which results in failure to salvage hypoxanthine and guanine to the corresponding nucleotides IMP and GMP.*
- *Inability to utilize PRPP in the salvage pathway leads to **PRPP accumulation,** which, in conjunction with low levels of IMP and GMP, causes chronic allosteric activation of PRPP glutamyl amidotransferase and **excessive purine synthesis.***
- *The excess purines are degraded to uric acid causing increased blood levels of this metabolite (**hyperuricemia**) and deposition of sodium urate crystals in the joints and kidneys.*
- *Patients with Lesch-Nyhan syndrome experience gout-like episodes of **joint pain** and **kidney stones** as well as severe neurologic problems, including **self-mutilation,** spastic movements, and mental retardation.*
- *Allopurinol treatment alleviates the symptoms of uric acid overproduction but does not remedy the neurologic problems.*

CLINICAL PROBLEMS

1. Severe combined immunodeficiency arises from inhibition of lymphocyte proliferation because B and T cells are particularly sensitive to allosteric inhibition of which of the following enzymes of purine nucleotide metabolism?

   A. Xanthine oxidase

   B. Dihydrofolate reductase

   C. Adenosine deaminase

   D. Ribonucleotide reductase

   E. Hypoxanthine-guanine phosphoribosyltransferase

A 68-year-old woman complains of chronic fatigue that has worsened over the past month. She has experienced recent bouts of nausea and diarrhea. History indicates that she changed her diet in order to lose some weight 3 months before she started experiencing these symptoms. Blood work reveals a macrocytic anemia. Neurologic examination is within normal limits.

2. These findings suggest a deficiency of which of the following vitamins?

   A. Niacin (B_3)

   B. Riboflavin (B_2)

   C. Vitamin B_{12}

   D. Folic acid

   E. Vitamin C

A 47-year-old man complains of pain in the joints of his left big toe, which are obviously swollen and tender. The pain has been chronic but became intolerable the day after Thanksgiving when he had a large meal and several glasses of red wine. He is obese, and his past medical history is significant for removal of kidney stones.

3. Which of the following is involved in the pathophysiology of this patient's condition?

   A. Elevated orotic acid

   B. Elevated uric acid

   C. Deficiency of folic acid

   D. Anemia

   E. Hypoglycemia

4. Methotrexate is a potent anticancer agent that starves dividing cells of deoxyribonucleotides through direct inhibition of which of the following enzymes?

   A. Ribonucleotide reductase

   B. Xanthine oxidase

   C. Carbamoyl phosphate synthetase II

   D. Thymidylate synthetase

E. Dihydrofolate reductase

F. Adenosine deaminase

A 2-year-old boy's mother is concerned about his tendency to bite himself to the point of bleeding. The boy's fingers show scarring and several scabs, and his lips are swollen and bruised. He exhibits poor coordination, poor muscle tone, and frequent jerking movements of his arms and legs. He is significantly delayed in speech. His urine is orange in color and "gritty."

5. Which of the following is the most likely diagnosis?

   A. Tay-Sachs disease

   B. Gout

   C. Lesch-Nyhan syndrome

   D. Severe combined immunodeficiency

   E. Cerebral palsy

ANSWERS

1. The answer is D. Impaired immune function in severe combined immunodeficiency (SCID) is the direct result of blocked DNA synthesis due to inadequate supplies of deoxyribonucleotides in B and T cells. This effect arises by dATP-induced allosteric inhibition of ribonucleotide reductase, which catalyzes reduction of the 2′-hydroxyl groups on ADP and GDP to form dADP and dGDP. The ultimate cause of many cases of SCID is adenosine deaminase deficiency, which leads to accumulation of dATP and consequent inhibition of ribonucleotide reductase. Although the other enzymes mentioned are also involved in purine nucleotide metabolism, their deficiencies do not lead to SCID.

2. The answer is D. Several vitamin deficiencies can cause anemia due to reduced DNA synthesis in the erythropoietic cells of the bone marrow, especially folic acid and vitamin B_{12} (cobalamin), which are particularly prevalent among elderly patients due to poor diet and reduced absorption. In addition, deficiencies of either folic acid or vitamin B_{12} could produce the megaloblastic anemia seen in this patient. However, the absence of neurologic symptoms, a hallmark of vitamin B_{12} deficiency, makes that diagnosis less likely than folic acid deficiency.

3. The answer is B. This patient shows many symptoms consistent with an episode of gout. His joint pain is due to gouty arthritis, an inflammatory condition arising from deposition of sodium urate crystals. The swelling in the joints of his big toe (tophaceous gout) is also a manifestation of this phenomenon. In his case, the episode seems to have been triggered by excessive eating at Thanksgiving dinner along with alcohol consumption, leading to degradation of large quantities of purine nucleotides and consequent increased flux through the pathway that produces uric acid. Whether his gout arises from impaired excretion of uric acid or is due to a mutation of PRPP synthetase

cannot be determined from the data. Analysis of his blood may confirm the gout if high concentrations of uric acid (hyperuricemia) are present.

4. The answer is E. Methotrexate is an analog of folic acid that binds with very high affinity to the substrate-binding site of dihydrofolate reductase, the enzyme that catalyzes conversion of DHF to THF, which is used in various forms by enzymes of both the purine and pyrimidine de novo synthetic pathways. Thus, synthesis of dTMP from dUMP catalyzed by thymidylate synthetase and several steps in purine synthesis catalyzed by formyltransferase are indirectly blocked by the action of methotrexate because both those enzymes require THF coenzymes.

5. The answer is C. This patient's self-mutilation behavior, neurologic symptoms, and developmental delay are all consistent with a diagnosis of Lesch-Nyhan syndrome. This disorder is due to deficiency of HGPRT, which prevents salvage of hypoxanthine and guanine to their respective nucleotides, IMP and GMP. This leads in turn to hyperactivity of the purine synthesis pathway, excessive purine degradation, and overproduction of uric acid. The gritty substance and orange color of the patient's urine are due to excretion of both dissolved uric acid and precipitated sodium urate. Gout might account for the excessive uric acid production but not the neurologic symptoms. Self-mutilation is not characteristic of either Tay-Sachs disease or cerebral palsy.

CHAPTER 11
NUCLEIC ACID STRUCTURE AND FUNCTION

I. Overview of Nucleic Acid Function

A. DNA is the **chemical basis of heredity.**

1. DNA comprises the genetic material by which the information to make proteins and **RNA** is stored and transmitted to offspring.

2. DNA is a linear polymer of deoxyribonucleotides in which the **sequence** of purine and pyrimidine bases encodes cellular RNA and protein molecules.

3. DNA is highly organized into **chromosomes,** structures that allow the DNA to be packaged tightly for storage in the **nucleus** of the cell.

   a. A diploid human cell contains 46 chromosomes within a **1 μm-diameter nucleus.**

   b. In order to fit such long molecules in a confined space, the **DNA must be compacted.**

B. The individual **genes** of defined sequence in the DNA specify or **encode** proteins and RNAs needed for all cellular functions.

1. **Replication,** the process by which copies of the DNA are made, must have very **high fidelity** or accuracy to ensure proper function of gene products and healthy offspring.

2. Most errors that occur during replication or as a result of oxidative or chemical damage, termed **mutations,** are **repaired** before cell division.

3. When these mutations are not repaired, a heritable change in the DNA occurs.

   a. Such a change may alter function or regulation of a gene product in daughter cells upon mitosis.

   b. If this occurs in germline cells, the mutation may be passed to offspring through gametes.

4. Mutations change the **sequence** of DNA bases and may thus lead to cellular dysfunction or disease.

   a. Proteins bearing amino acid substitutions may not attain their correct conformations or properly serve their physiologic functions.

   b. Mutations in regulatory areas near genes can prevent proper control over their expression.

C. RNA molecules operate at critical points in many of the processes that involve expression of the information represented in the DNA.

1. **Transcription** is the process by which RNA copies of the genes are synthesized as the first step leading to **gene expression.**

2. **Messenger RNAs (mRNAs)** carry copies of the genes that can be **translated into proteins** (see Chapter 12).
3. Other specialized RNAs, ribosomal RNA (rRNA), transfer RNAs (tRNAs), and small RNA molecules are not translated into protein but have central roles in gene expression and protein synthesis.
4. In eukaryotes, mRNAs are initially transcribed as **heterogeneous nuclear RNA,** which still contains **intervening sequences** of the gene and **must undergo processing** to attain the final mRNA structure.

II. Structure of Chromosomal DNA

A. The information contained in the DNA is represented by the **sequence** of the bases of the polymer, the purines **adenine (A)** and **guanine (G)** and the pyrimidines **cytosine (C)** and **thymine (T).**
   1. The deoxyribonucleotides in the DNA polymer are connected by **phosphodiester bonds** between the 5′-phosphate group attached to one deoxyribose sugar and the 3′-hydroxyl group of the next sugar.
   2. This **sugar-phosphate backbone** is located on the outside of the structure.

B. **DNA is double-stranded.**
   1. The two strands run in reverse polarity or **antiparallel** to each other.
   2. The strands are held together by **hydrogen bonding** between the bases, with a purine always bonded with a pyrimidine in specific **base pairs.**
      a. Base pairs have a preferred structural **complementarity.**
      b. A pairs with T via two hydrogen bonds.
      c. G pairs with C via three hydrogen bonds.
   3. The entire structure is twisted along the main axis in the form of a right-handed **double helix.**
      a. In **B DNA,** the form that occurs most commonly under physiologic conditions, the helix has **major and minor grooves,** which provide access for **protein binding** to the DNA.
      b. The double helix is also stabilized by hydrophobic **base stacking interactions** in the core.
   4. Cooperation between the many hydrogen bonds and base-stacking interactions makes DNA very stable to chemical treatments.
      a. Increased heat, decreased salt concentration, or extremes of pH can force the DNA duplex **to melt open** (or "unzip") by disrupting the hydrogen bonds between the strands in a process called **denaturation.**
      b. The **melting point (T_m)** at which the DNA duplex is half-unzipped depends on length of the DNA and its percentage of G and C bases, which provide a stronger base-pair interaction than A and T pairs.
      c. Nevertheless, the two strands of DNA must be separated by proteins under physiologic conditions for important processes like **replication** and **transcription.**
   5. **Deoxyribonucleases** (DNAses) catalyze hydrolysis of phosphodiester bonds in the DNA backbone, leaving a 5′-phosphate and a 3′-hydroxyl on the ends.
      a. **Endonucleases cleave** within the **interior** of a DNA strand, as in the case of **restriction endonucleases,** which cleave at sites having **specific sequences.**
      b. **Exonucleases cleave** the last nucleotide from either the **3′** or **5′ ends** of the DNA strand, depending on the enzyme's specificity.

C. **Compacting** of the DNA for storage in the limited space available in the cell's nucleus is accomplished by **binding of proteins** (Figure 11–1).
  1. A family of small, basic proteins called **histones** is responsible for major interactions with DNA in the formation of **nucleosomes.**
     a. Prokaryotic histones are of five types: H1, H2A, H2B, H3 and H4; vertebrates also have histone H5.
     b. Binding of the positively charged histones to DNA **neutralizes negative charges** of the phosphate groups along the DNA backbone, which allows the **DNA to bend** much more easily than naked DNA.

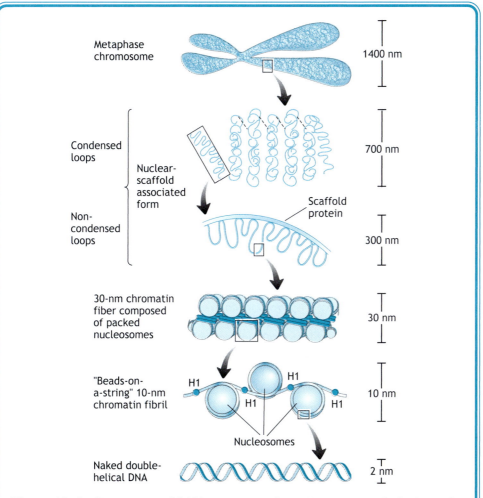

Figure 11–1. Compaction of DNA into a metaphase chromosome. Each phase of condensation or compaction reduces availability of DNA for replication or transcription to the point where a metaphase chromosome is nearly inert for these processes. The overall, packaging reduces the length of the DNA 10^4-fold.

2. Two molecules each of the similarly sized histones H2A, H2B, H3, and H4 bind together in an **octamer** that forms the core of a **nucleosome.**
   a. DNA wraps around each nucleosome core 1.75 times, so that the nucleosomes form at uniform intervals along the DNA.
   b. Approximately 146 base pairs of DNA are involved in forming each nucleosome, with about 30-base-pair **linker regions** between them.
   c. Various types of **histone H1 (or H5)** bind loosely to the linker regions to help organize the nucleosomes into higher-order structures.
D. **Tertiary** and **quaternary structures** of DNA allow even further condensation of nucleosome-coated DNA into the highly compacted structure of the **chromosome** (Figure 11–1).
   1. **Nucleofilaments** are organized by coiling the nucleosome-coated DNA into 30-nm fibers that also contain non-histone proteins and some RNA molecules.
   2. Packing of the nucleofilaments around **scaffold proteins** is the final level of condensation into **chromosomes.**
   3. Two copies of each chromosome are stored in the nucleus and take on a special structure during **metaphase of mitosis** prior to **cell division.**
      a. The two chromosomes undergo further compaction into **chromatids,** the familiar **dense structures** visible when stained and viewed under the microscope.
      b. **Two sister chromatids** are connected by a **centromere** composed of proteins bound to an A+T-rich region of the DNA.
      c. The centromere forms the point of attachment for the **mitotic spindle.**

III. Replication

A. In order that a complete complement of the genetic material may be inherited by daughter cells during cell division or by offspring from parents, the DNA must be copied with **high fidelity** by a process called **DNA replication.**
   1. The DNA region to be replicated is copied by what is referred to as a **semiconservative** mechanism.
      a. The DNA region must be **opened up** from its double-stranded state to two halves of complementary, single-stranded DNA (ssDNA).
      b. Each of the strands then serves as the **template** for synthesis of a new complementary, daughter strand.
   2. The two new daughter strands are formed along the template by base pairing with deoxynucleotide triphosphates (**dNTPs**) serving as the **building blocks.**
B. Prokaryotic DNA replication is accomplished by **DNA polymerases,** large multienzyme complexes that move out bidirectionally from the origin of replication.
   1. DNA replication begins with protein binding to the **origin of replication,** a unique sequence in the bacterial chromosome, causing a short region of double-stranded DNA (dsDNA) to unwind (Figure 11–2).
   2. **Single-stranded DNA binding proteins (SSBs)** bind to this short region of ssDNA, which promotes further unwinding of a nearby A+T-rich region.
   3. DNA **helicase,** an enzyme that catalyzes **DNA unwinding,** can then bind to this region and begin working in cooperation with SSBs.
      a. Helicase forces open dsDNA by breaking hydrogen bonds stabilizing the base pairs ahead of the moving replication fork.
      b. SSBs then bind to the ssDNA to prevent reannealing to the double-stranded state.

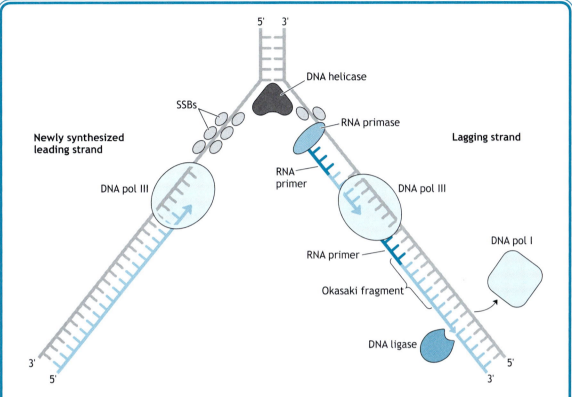

Figure 11–2. The prokaryotic DNA replication fork. A schematic representation of semi-conservative replication of DNA by different mechanisms on the leading and lagging strands by DNA polymerase III (DNA pol III) is shown. Other enzymes and accessory proteins that participate in initiation, elongation, and ligation phases of the process are indicated, with DNA pol I depicted as having just dissociated from a completed Okasaki fragment. SSBs, single-stranded DNA binding proteins.

4. Once the **replication fork** is established, other proteins begin assembling the functional **DNA replication complex.**
   a. DNA polymerases are able to copy the single-strand **template** DNA by operating only in a **5′ to 3′ direction.**
   b. This limitation presents a problem to copy DNA of the unzipped, antiparallel strands in opposing directions, both toward and away from the replication fork.
   c. To solve this problem, different mechanisms are used to make dsDNA using the single-stranded templates on the forward (**leading**) strand and the retrograde (**lagging**) strand (Figure 11–2).
5. **DNA polymerase III (DNA pol III),** the main DNA polymerase of *Escherichia coli*, synthesizes DNA continuously on the leading strand and discontinuously on the lagging strand.
   a. DNA polymerases require that a **primer** with an available 3′-hydroxyl end be annealed to the template.

b. The **nascent** or growing polynucleotide chain being made as complement to the leading strand continuously provides a 3′ end that is extended by DNA pol III.
 6. Replication on the **lagging strand** is **discontinuous** because polymerases can only copy the single-stranded region available at the fork and only in the 5′ to 3′ direction.
   a. A short **RNA primer** is first synthesized nearest the 3′ end of the fork by **primase**, which is actually a DNA-directed RNA polymerase.
   b. DNA pol III binds to the primer-template end and extends the primer by adding deoxyribonucleotides during the **elongation** step.
   c. Short pieces of DNA called **Okasaki fragments** are made in this way and each fragment is completed when DNA pol III bumps up against the primer end of the previous fragment (Figure 11–2).
   d. The RNA primers are excised and simultaneously replaced with DNA by **DNA pol I,** which also has **5′ to 3′ exonuclease** activity.
   e. **DNA ligase** then seals the remaining nick by catalyzing formation of a phosphodiester bond with ATP as energy donor.

INHIBITORS OF DNA REPLICATION AS ANTICANCER AND ANTIVIRAL AGENTS

- *When **nucleoside analogs,** such as cytosine arabinoside (AraC), azidothymidine (zidovudine or AZT), and dideoxyinosine (ddI), are converted into the corresponding nucleotides by salvage pathways, they can be incorporated into nascent DNA strands by DNA polymerases.*
- *These compounds have **modified sugars** that are not capable of forming downstream phosphodiester bonds, which **blocks further elongation** of the chains.*
- *Although these drugs effectively inhibit the replication of DNA in all cells, they are highly toxic to rapidly proliferating cells, such as cancer cells and cells infected by virus.*

 C. **Topoisomerases** are responsible for relieving **supercoils** in the dsDNA that occur by twisting and fold-back as the DNA is unwound ahead of the replication fork.
   1. If **supercoils** or **superhelices** were not removed, they would eventually **block** movement of the replication fork by preventing further DNA unwinding.
   2. Topoisomerases are ATP-dependent enzyme complexes that bind to and **relax the supercoiled regions** of DNA.
     a. **Type I topoisomerases** bind the dsDNA region, **cut one strand,** and allow **controlled rotation** around the intact strand causing the over-twisted DNA to relax.
     b. **Type II topoisomerases** bind to two double-stranded sides of a DNA superhelical loop, **make a double-stranded cut** on one side, and allow the intact DNA segment to pass through the break to relax the over-twisted DNA.
   3. The severed phosphodiester bonds are then reconnected by the ligase activity of the topoisomerase.

TOPOISOMERASE INHIBITORS AS ANTICANCER AND ANTIBIOTIC AGENTS

- *The anticancer agents **etoposide** and **amsacrine** are **inhibitors of topoisomerase II.***
- ***Camptothecin,** an **inhibitor of topoisomerase I,** is an effective **anticancer agent** that converts the enzyme to become a DNA-damaging agent.*

- Bacterial topoisomerases (called **DNA gyrases**) are inhibited by several important classes of antibiotics, including the coumarins, such as **novobiocin;** quinolones, such as **nalidixic acid;** and fluoroquinolones, such as **ciprofloxacin.**

D. DNA replication is regulated as a balance between high speed and efficiency (**processivity**) and the need for high fidelity.
1. DNA pol III is **processive** because it can add thousands of nucleotides to the nascent strand before falling off the template.
2. **Fidelity** of match between the template and the newly synthesized copy is maintained at a high level by enzymes with **proofreading** activity.
    a. DNA pol III makes occasional errors by incorporating an incorrect nucleotide to create a base-pair **mismatch** at a frequency of 1 per 10,000 nucleotides.
    b. Mismatches are corrected by **proofreading, 3′ to 5′ exonuclease** activities associated both with DNA pol III and DNA pol I, which recognize and excise the mismatched nucleotides.
    c. The polymerase activities then replace the missing nucleotides with correct matches.
    d. These mechanisms reduce the overall error rate to 1 mismatch per 10^{10} nucleotides.

E. Eukaryotic DNA replication is similar to that of prokaryotes but more complex in scale, and the process is coordinated with the **cell cycle.**
1. Compared with the process in bacteria, replication of DNA of a human cell requires **multiple origins of replication,** each of which leads to copying of **replicons,** regions 30 to 300 kilobase pairs in size.
2. DNA replication occurs during the synthetic or **S phase** of the cell cycle in preparation for mitosis.
3. **Slipped mispairing** at the replication fork can cause repeated copying of some sequences within the tract and thus lead to **expansion of trinucleotide repeat (TNR)** tracts at the 5′ ends of certain genes.
    a. TNR expansion interferes with transcription of the mRNA or, if the tract is in the coding region, produces a mutant, defective protein.
    b. This mechanism is responsible for a group of diseases called TNR disorders.

TRINUCLEOTIDE REPEAT DISORDERS

- A group of over a dozen **inherited neurologic diseases** exhibits genetic **instability** due to dynamic mutation that shows **anticipation,** a genetic phenomenon whereby affected offspring in successive generations show symptoms earlier and of a more severe nature than their parents.
- **Huntington disease** is an **autosomal dominant** disorder involving degeneration of the striatum and cortex that manifests as **motor dysfunction** in midlife and leads to progressive **loss of cognitive function** and death.
    – The gene responsible for Huntington disease has a **CAG repeat tract** coding for **polyglutamine** at the N-terminal end of the protein huntingtin, the function of which is impaired when the tract exceeds 35 repeats.
    – Anticipation occurs in Huntington disease as the TNR tract expands in length from one generation to the next, causing progressively greater interference with the protein's function.
- **Fragile X syndrome** is an X-linked disorder arising from inactivation of **FMR1,** a gene that encodes a protein critical for **synaptic function.**

- FMR1 *has a* **CGG repeat tract** *in the 5′ untranslated region; when the length of the tract expands beyond 200 copies, the* FMR1 *promoter becomes extensively* **methylated** *and is thereby inactivated (the* **threshold** *effect).*
  - *Fragile X syndrome is the most common inherited form of* **mental retardation,** *with a frequency of 1 in 4000 males and 1 in 8000 females.*
  - *Symptoms of Fragile X syndrome include cognitive impairment, autism, seizures, and hyperactivity.*

    F. Humans and other eukaryotes have **linear chromosomes,** which create special problems for replication of DNA at the chromosome ends.

      1. The **chromosomes become shorter** at each round of DNA replication after removal of the RNA primer from the lagging strand.

      2. To minimize the possibility that shortening might delete important gene regions, the **chromosome ends** are formed of **telomeres.**

        a. Telomeres are regions of DNA that do not contain any genes and in humans consist of multiple **repeats** of the sequence 5′-TTAGGG-3′ that may be up to 10 kilobase pairs long.

        b. The end DNA **loops back** to form a duplex that is stabilized by telomere binding proteins.

        c. In normal, **aging** cells, telomeres shorten at each round of DNA replication, eventually leading to their complete removal; subsequent rounds of replication erode portions of essential genes, producing **cell cycle inhibition** and replicative cell **senescence.**

      3. In **germ-line cells** and other cell types that do not undergo aging, telomere lengths are maintained by **telomerases.**

        a. Telomerases can **bind to the single-stranded 3′ end** of the chromosome after DNA replication and **extend** it by adding new repeat elements.

        b. After extension of the end by telomerase, DNA polymerases can prime and copy the region.

TELOMERASE ACTIVITY IS HIGH IN CANCER

- **Senescence** *by regulation of telomere length is considered an important* **safeguard** *against uncontrolled proliferation of somatic cells.*
- *Most human cells have very low telomerase activity, but* **cancer** *cells have* **high telomerase** *activity, which allows them to avoid senescence and become "immortal."*
- *Telomerase inhibitors are under development as potential anticancer agents.*

IV. Mutations and DNA Repair

    A. **Mutations** or heritable alterations in the DNA sequence that affect protein structure or gene expression can occur in many ways and may be passed to daughter cells during cell division.

      1. Errors in DNA replication can produce a variety of mutations by failure of proofreading mechanisms.

      2. **Point mutations** or single base substitutions are classified as transitions or transversions.

        a. **Transitions** are defined as the substitution of one purine for another on the same strand (eg, A to G or G to A); likewise for pyrimidine substitutions.

        b. **Transversions** are defined as the substitution of a purine for a pyrimidine or vice versa (eg, A to C or T to G).

B. **Chemical modification of DNA** caused by environmental **mutagens** may lead to changes in the function or expression of genes.
  1. Chemical reactions can modify DNA bases leading to **altered base pairing** in subsequent rounds of replication.
      a. **Alkylating agents** are compounds that are metabolized within cells to unstable species that react with sites on the DNA bases, which may alter their base-pairing properties and eventually cause mutations.
      b. Some compounds react with bases to produce **adducts,** which are covalently modified bases that are spontaneously ejected from the DNA. The **abasic site** formed as a result cannot base-pair properly upon replication.
  2. **Intercalating agents** are aromatic compounds that fit between the base pairs in the core of DNA structure and lead to **insertions and deletions** of one or more base pairs upon replication.
  3. **Ultraviolet light** causes neighboring thymine bases to form **thymine dimers** that block replication and gene expression.

CHEMICAL CARCINOGENESIS: MUTATION OF DNA LEADING TO CANCER

- *Cigarette smoke* contains **aryl hydrocarbons** such as **benzo[a]pyrene** that, once metabolized to reactive compounds, can form alkyl **adducts of DNA bases** leading to mutations and cancers of the lung and many other organs.
- *Smoked and grilled foods* are coated with **nitrosamines,** which can alkylate any of the bases of DNA but particularly guanine to cause cancers of the digestive tract and other organs.
- The **UV-B** component of ultraviolet light in **sunlight** can damage DNA by forming thymine dimers and is a major contributor to **skin cancer.**
- Ionizing radiation, such as **gamma rays and x-rays,** causes complex types of DNA damage that are difficult to repair, including **double-strand and single-strand breaks** and **cross-links** that may lead to leukemia and cancers of many organs.

  C. Many types of DNA damage can be repaired by specialized enzyme systems.
  1. **Base excision repair** involves the removal of abnormally modified bases by **glycosylases** with subsequent replacement with the appropriate base.
  2. **Nucleotide excision repair** involves the removal of the region surrounding a modified base or single-strand break by nuclease-mediated excision (cutting) of the DNA strand on either side of the lesion followed by filling of the resulting gap.
  3. **Mismatch repair** involves elements of both base-excision and nucleotide excision mechanisms.
  4. Repair of **double-strand breaks** requires multi-enzyme mechanisms, but repair may be imperfect with retention of some mutated sequences.

XERODERMA PIGMENTOSUM

- *Xeroderma pigmentosum is caused by a* **defect in excision repair** *of thymine dimers, most frequently due to the absence of a* **UV-specific excinuclease,** *an enzyme that helps remove thymine dimers.*
- This is a rare, **autosomal recessive** disorder characterized by **extreme sensitivity to sunlight.**
- During their first two decades, patients suffer dramatic **changes in the skin,** including excessive dryness, pigmentation, atrophy, and **hyperkeratosis** (thickened precancerous outgrowths of the dermis), with **eye** manifestations such as corneal cloudiness or ulceration.
- Patients with xeroderma pigmentosum are prone to develop **skin cancer** later in life.

FANCONI ANEMIA

CLINICAL
CORRELATION

- *Fanconi anemia arises from a decreased ability to repair **interstrand DNA cross-links.***
- ***Defective DNA repair** leads to severe clinical manifestations in this congenital **autosomal recessive** disorder.*
- *Patients exhibit microcephaly with **mental retardation,** bone marrow insufficiency leading to **anemia** and **leukopenia** (decreased WBC count), and hypoplastic kidneys.*
- *Affected **children** are hypersensitive to DNA-damaging agents and prone to a variety of **cancers** early in life.*

V. RNA Structure

   A. All **RNA molecules represent copies of genes** on the cellular DNA, but there are some important differences in structure between DNA and RNA.

      1. The features of RNA structure that distinguish it from DNA follow:

         a. Presence of **ribose** as the **sugar** in the backbone of RNA rather than 2′-deoxyribose as in DNA.

         b. Thymine (T) in DNA is replaced by **uracil (U)** in RNA.

         c. RNA is a **single-stranded** version of one strand of the DNA sequence, at least as initially synthesized.

         d. RNA can form complex, variable **secondary structures** by internal foldback and **intramolecular base pairing** between complementary regions of the molecule.

      2. Most types of cellular RNA are involved in various steps in **protein synthesis** or **gene expression.**

   B. The function of the ribosome, including its main catalytic activity, depends on several forms of **ribosomal RNA (rRNA).**

      1. Ribosomes are large **nucleoprotein machines** composed of large and small subunits that carry out **protein synthesis.**

      2. Prokaryotic ribosomes contain three rRNAs: 16S rRNA in the small (30S) subunit and 23S and 5S rRNA molecules in the large (50S) subunit.

      3. Eukaryotic ribosomes contain four rRNAs analogous to those in prokaryotes: the 18S rRNA of the small (40S) subunit and the 28S, 5.8S, and 5S of the large (60S) subunit.

      4. Cells have many ribosomes, so rRNAs comprise the majority (~80%) of cellular RNA.

   C. mRNA represents an RNA copy of a gene, which directs synthesis of a specific protein by the ribosomes.

      1. Prokaryotic genes encode protein sequences directly with no intervening non-coding DNA, so that mRNA transcripts serve as direct templates for protein synthesis.

      2. In eukaryotes, the first step in mRNA synthesis is transcription of the template or "non-coding" strand of DNA into a large **heterogeneous nuclear RNA (hnRNA),** which undergoes **processing** to remove intervening, non-coding sequences (**introns**) and to add **stabilizing structures.**

   D. tRNAs are small molecules that function as adaptors to convert or **translate** the **nucleotide** sequence information of mRNAs into the **amino acid** sequences of the proteins they encode.

      1. Many different forms of tRNA occur in cells, at least 1 for each of the 20 common amino acids.

2. The tRNAs are 65–110 nucleotides long and their backbones fold back to allow for **intramolecular hydrogen binding** (base pairing or hybridization) to form a **cloverleaf secondary structure.**

3. Base stacking effects and some unusual forms of hydrogen bonding between the bases cause tRNAs to take on a **tertiary structure** that is roughly **L-shaped.**

   a. The 3′ OH end of all tRNAs has the same sequence, 5′-CCA-3′, forming the **acceptor stem** to which a specific **amino acid attaches.**

   b. At the opposite side of the molecule, is the **anticodon loop,** containing the **3-base sequence** or **anticodon** that base pairs with the **codon,** or amino acid-specifying unit, of the mRNA (see Chapter 12).

   c. Other loops such as the **TψC loop** and **DHU loop** help the tRNA bind to various enzymes and to ribosomes.

4. The tRNAs undergo post-transcriptional modification to produce **specialized bases,** such as pseudouridine, dehydrouridine, and methylcytosine.

E. **Small nuclear RNA (snRNA)** molecules are components of **splicesomes,** which are complex nucleoprotein assemblies that process or **splice hnRNAs to mRNAs.**

VI. Transcription

A. **Transcription** is the process by which the template strand of DNA is copied into RNA for purposes of gene expression.

B. **DNA-dependent RNA polymerase** copies the sequence of the DNA template into a complementary RNA or **transcript.**

1. Like DNA polymerases, prokaryotic RNA polymerase (RNA pol) is a multi-protein complex that operates only in the 5′ to 3′ direction as it copies the template.

   a. The RNA pol **holoenzyme** has five subunits in its $\alpha_2\beta\beta'\sigma$ **complex.**

   b. The sigma factor, σ, can dissociate from the holoenzyme, leaving behind the **core enzyme,** which has the main catalytic activities.

2. The mechanism of transcription is identical for all forms of RNA and occurs in **multiple steps.**

3. To **initiate** transcription, the RNA pol holoenzyme binds to and slides (scans) along the DNA searching for an appropriate **promoter,** a specific sequence element that indicates the 5′ end of a gene.

   a. The **σ factor** of the holoenzyme binds to the DNA sequence 5′-TATAAT-3′, called the **TATA box,** within the promoter region guiding the holoenzyme to the site.

   b. RNA pol holoenzyme **unwinds** 17 base pairs of DNA to form the **pre-initiation complex.**

   c. RNA pol then forms the first phosphodiester bond between two base-paired ribonucleotides to **initiate** the new chain, in the **absence of a primer.**

   d. Once the first phosphodiester bond is formed, **σ factor** dissociates, which decreases the affinity of RNA pol for the promoter and allows the core enzyme to continue synthesis along the DNA.

4. **Elongation** of the transcript occurs by incorporation of ribonucleotides to create a copy or **RNA complement** of the DNA template.

   a. The RNA pol holoenzyme, the unwound portion of the template and the nascent RNA chain form the **transcription bubble,** which moves along the DNA during transcription (Figure 11–3).

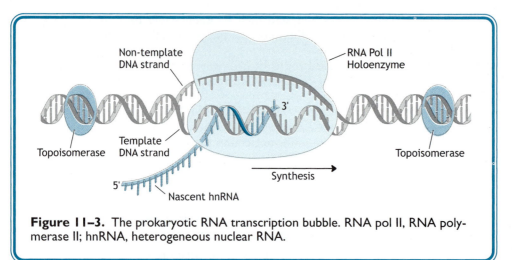

Figure 11–3. The prokaryotic RNA transcription bubble. RNA pol II, RNA polymerase II; hnRNA, heterogeneous nuclear RNA.

      b. Ribonucleotides are added to the nascent chain according to base-pairing rules, with C hydrogen bonding with G as usual and **U pairing with A of the DNA** and A pairing with T of the DNA.

      c. **Topoisomerases** prevent supercoiling ahead of and behind the moving bubble.

      d. **RNA pol** does not have nuclease activity, so it is **not capable of proofreading** and is more error-prone than DNA polymerase.

   5. Termination of transcription occurs when RNA pol traverses a termination signal, and this process may require the cooperation of ρ (rho) factor.

C. Eukaryotic transcription is more complex than in prokaryotes, mainly in terms of the nature of the **RNA polymerases,** the assembly of the **pre-initiation complex,** and the need for **processing** eukaryotic RNAs.

   1. Three DNA-dependent RNA polymerases operate in the transcription of eukaryotic genes.

      a. **RNA pol I transcribes the 28S, 18S, and 5.8S rRNA genes,** an activity that is localized to the **nucleolus,** a region of high nucleoprotein density in the cell's nucleus.

      b. **RNA pol II** is responsible for transcription of **snRNA** genes and of structural genes encoding **mRNAs** leading to protein synthesis.

      c. **RNA pol III** transcribes the **tRNA** genes and the 5S rRNA gene.

   2. General transcription factors (GTFs) that bind to eukaryotic promoters are functionally analogous to σ factor in prokaryotes.

      a. **TATA binding protein (TBP)** recognizes the **TATA box** element of the promoter on type II genes (those transcribed by RNA pol II), binds to it in a sequence-specific manner, and recruits other GTFs to form a complex.

      b. RNA pol II is then attracted to the complex to form the pre-initiation complex.

      c. Besides TBP, the GTFs and more specific transcription factors that regulate transcription of the many type II genes differ depending on the gene (see Chapter 12).

MUSHROOM TOXIN INHIBITS RNA POLYMERASE II

- *Each year, more than 100 people worldwide die after eating poisonous mushrooms.*
- *Ingestion of as little as 3 g of the **death cap mushroom** Amanita phalloides may constitute a lethal dose for some people.*
- *This mushroom produces the toxin, **α-amanitin,** a cyclic octapeptide having several modified amino acids and a central purine, which strongly binds to and **inhibits RNA pol II** and thereby **blocks elongation.***
- *RNA pol II is essential for proper function of cells in all tissues and organs, but potentially fatal **liver and kidney failure** is the main risk for victims of α-amanitin poisoning.*

   3. Removal of **introns** from hnRNA to leave only the **exons** or gene regions involved in directing protein synthesis in the finished mRNA is accomplished
   within the nucleus by **processing on spliceosomes** (Figure 11–4).

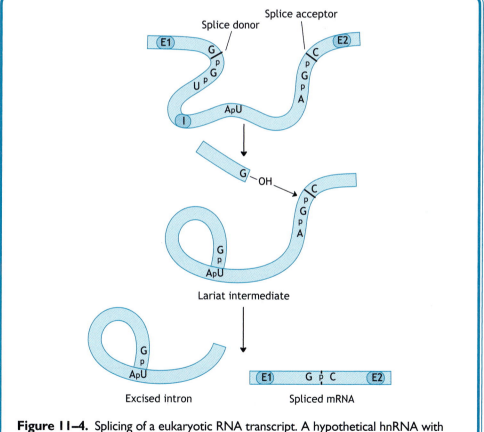

Figure 11–4. Splicing of a eukaryotic RNA transcript. A hypothetical hnRNA with two exons (E1 and E2) and a single, large intron (I) is shown. Splicing can be divided into two main reactions: initial attack of ribose near an A residue within the intron on the splice donor followed by attack of the newly available 3′ end of exon 1 (E1) on the 5′ end of exon 2 (E2) with coincident release of the intron. Special sequences surround the splice donor and acceptor sites. All steps occur within the spliceosome complex.

a. **Introns** of structural genes vary widely in size and sequence, but they tend to have **common sequences** at the intron:exon boundaries or **splice junctions.**

b. **Spliceosomes** are **nucleoprotein complexes** containing over 60 proteins and 5 snRNAs, which act to **position** and **coordinate the splicing** reactions that remove introns from the hnRNAs.

c. Splicing begins by **reaction** of an A base near the 3′ end of the intron with the 5′ end, which is cleaved in the process.

d. The **cleaved 5′ intron** end is tethered to the original A by a looped or **lariat** structure in a unique **5′ to 2′ phosphodiester** linkage between the backbone ribose sugars.

e. The 3′ OH end of the first exon then reacts with the 5′ end of the second exon with simultaneous cleavage to release the lariat and join the exons.

f. The most noteworthy aspect of the splicing reactions is the occurrence of **catalysis by the RNA** itself.

4. Most eukaryotic mRNAs have a **7-methylguanine cap** at the **5′ end,** which promotes **efficient translation** of the message and **protects it from degradation** by 5′ to 3′ exonucleases.

5. Most eukaryotic mRNAs end approximately 20 nucleotides downstream of the sequence, **AAUAA,** which permits addition of a **polyA tail** that **protects** the message from cleavage by 3′ to 5′ exonucleases.

CLINICAL PROBLEMS

A 5-year-old boy has a rough, raised lesion on his neck. Physical examination shows that he has excessive freckling and some erythema (redness) of his face, lips, neck, and upper extremities as well as some clouding of his corneas. His mother reports that he has a tendency to sunburn easily and has an aversion to direct sunlight. Pathologic evaluation of a biopsy of the lesion reveals it to be a malignant melanoma.

1. This patient most likely suffers from deficiency of an enzyme involved in the repair of which type of DNA damage?

   A. Base adducts

   B. Thymine dimers

   C. Abasic sites

   D. Mismatches

   E. Double-stranded breaks

   F. Single-stranded breaks

2. In this case, which repair mechanism is most likely defective?

   A. Base excision repair

   B. Mismatch repair

C. Nucleotide excision repair

D. 5′ to 3′ exonuclease

E. 3′ to 5′ exonuclease

Sickle hemoglobin (HbS) differs from normal adult hemoglobin (HbA) at amino acid number 6 of the β-globin chain, where HbS has a Val and HbA a Glu.

3. This amino acid substitution arose from what type of mutation?

A. Missense

B. Nonsense

C. Insertion

D. Deletion

E. Amplification

A 37-year-old man reports suffering from nausea, vomiting, and mild abdominal pain over the past 7 hours, ever since he returned from a hike in the woods during which he had picked and eaten some wild mushrooms.

4. His symptoms most likely arise from toxin-induced inhibition of which of the following enzymes?

A. Topoisomerase

B. DNA polymerase

C. Helicase

D. RNA polymerase I

E. RNA polymerase II

F. Telomerase

5. Cancer cells avoid replicative senescence by maintaining integrity of their chromosome ends through increased activity of which of the following enzymes?

A. Topoisomerase

B. DNA polymerase

C. Helicase

D. RNA polymerase I

E. RNA polymerase II

F. Telomerase

A 7-year-old boy is referred by his school nurse for evaluation of hyperactivity accompanied by developmental delays in speech and motor skills. The nurse is concerned about his IQ tests, which indicate mild mental retardation. Family history indicates that his mother and maternal aunt both have learning disabilities and one of his maternal uncles lives in a group home for the mentally retarded. Physical examination shows that the boy is normocephalic and normally pigmented.

6. Analysis of a sample of this patient's DNA for genetic abnormalities should focus on which of the following genes?

A. *FMR1* (Fragile X)

B. *XP-A* (Xeroderma pigmentosum gene)

C. *HD* (Huntington disease)

D. *FANC* genes (Fanconi anemia)

E. *GALC* (galactosylcerebrosidase, Krabbe disease)

ANSWERS

1. The answer is B. The patient has many of the features characteristic of xeroderma pigmentosum. The lesion, hyperpigmentation (freckles), and erythema are located over sun-exposed areas. His corneas have also suffered damage from exposure to ultraviolet irradiation from the sun. His photosensitivity is also manifested in easy sun-burning and aversion to sun exposure. This condition often leads to skin cancer.

2. The answer is C. Thymine dimers are repaired by the process of nucleotide excision repair, which involves many enzyme activities that recognize the mutated structure, cut the DNA strand on both sides of the mutation, remove (excise) the affected fragment, and then refill the gap. One of the major genes leading to xeroderma pigmentosoum encodes a specific excinuclease.

3. The answer is A. Each amino acid in a protein is specified by a 3-base or triplet sequence on the mRNA (see Chapter 12). A missense mutation occurs when one or more of the bases in the triplet are changed so that a different amino acid is specified. The protein is still produced but may be defective, as in the case of sickle hemoglobin, where the replacement of the polar glutamate (Glu) to a nonpolar valine (Val) makes the protein "sticky" and gives it a tendency to form polymers in the deoxyhemoglobin state under conditions of low PO_2. Nonsense mutations are those in which the base change creates a stop codon that does not specify an amino acid but instead causes termination of the protein. Insertions, deletions, and amplifications are more likely to cause synthesis of grossly defective or truncated proteins.

4. The answer is E. The patient's history of sudden onset of mild gastrointestinal symptoms after eating hand-picked wild mushrooms suggests poisoning by α-amanitin, a potent, selective inhibitor of RNA pol II, the critical enzyme for transcription of structural genes in human cells. There is no treatment for α-amanitin poisoning beyond palliative care, and if sufficient toxin has been ingested, death due to liver failure is a possible outcome.

5. The answer is F. The ends of linear chromosomes cannot be replicated by normal cells due to the inability to prime and synthesize Okasaki fragments on the lagging strand for replication by the human equivalent of DNA pol III. As cells divide repeatedly for tissue maintenance, the chromosome ends containing telomeric sequences eventually dwindle to the point where there is loss of genetic material encompassing structural genes. This leads to replicative senescence. To avoid this aging condition, cancer cells activate expression of telomerase, an enzyme that has a built-in RNA primer and the polymerase activity needed to make multiple copies of the six-base repetitive sequence of the telomeres.

6. The answer is A. The family history in this case strongly points toward Fragile X syndrome as the most likely diagnosis, which would be indicated if a mutation was discovered in the X-linked gene *FMR1*. Fragile X syndrome is a trinucleotide repeat disorder characterized by mental retardation. If confirmed, the patient's symptoms appear to be more severe than the affected individuals of the previous generation indicative of anticipation. The findings that he is normocephalic and normally pigmented are inconsistent with Fanconi anemia. Early onset in life and lack of motor impairment are inconsistent with Huntington disease. Krabbe disease is ruled out due to the lack of motor impairment, seizures, deafness, or blindness.

CHAPTER 12
GENE EXPRESSION

I. The Genetic Code

A. Conversion of the information present in **genes into proteins** of proper structure and function is accomplished by a series of highly regulated processes collectively termed **gene expression.**

1. The first process, **transcription** of the DNA sequences of the genes into messenger RNA (mRNA), has been discussed in Chapter 11.
2. The process by which the linear sequence of nucleotides in the mRNA is converted into protein sequence is called **translation.**

B. Translation depends on the **genetic code.**

1. The genetic code is a set of **"words"** or **codons** that are read as the nucleotide sequence of the mRNA and translated into the protein sequence it specifies.
2. The genetic code has **64 words.**
3. The words in the mRNA are **contiguous,** ie, punctuation-free.
4. Each codon is a **triplet** of three nucleotides and is **unambiguous,** specifying only a single amino acid.
5. The code is **degenerate** in that some amino acids are specified by more than one codon.

C. The codons of the mRNA cannot directly recognize the amino acids they specify, but this function is served by transfer RNAs (**tRNAs**) that act as **adaptors** to match each codon to its corresponding amino acid.

1. A tRNA molecule displays a three-base **anticodon** that is **complementary to the codon** on one end of its folded structure and carries the corresponding amino acid on the opposite end (Figure 12–1).
2. There are not 64 different tRNAs, one for each codon, but instead the tRNAs are capable of unconventional base pairing (**"wobble"**) with the codons during translation of the mRNA.
3. The three **termination or stop codons,** UAA, UAG, and UGA, do not specify amino acids and thus do not base pair with specific tRNAs.
4. The **code is not entirely universal;** there are minor differences between the codes used for synthesis of proteins encoded by **nuclear versus mitochondrial genes** of human cells.

II. Steps in Translation

A. Covalent **coupling of amino acids to their tRNAs** is a **high-fidelity** process mediated by very specific enzymes.

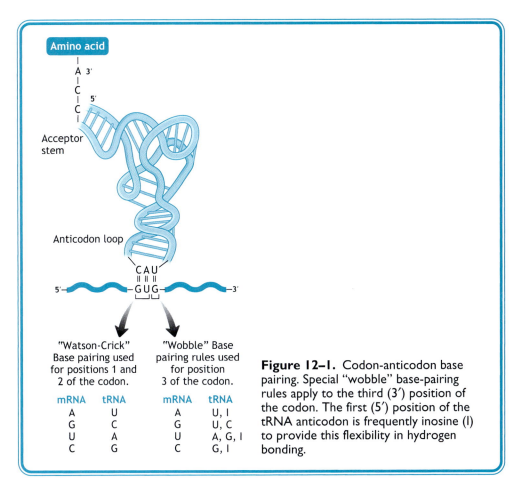

Figure 12–1. Codon-anticodon base pairing. Special "wobble" base-pairing rules apply to the third (3′) position of the codon. The first (5′) position of the tRNA anticodon is frequently inosine (I) to provide this flexibility in hydrogen bonding.

1. **Aminoacyl-tRNA synthetases** are the critically important enzymes responsible for coupling (**charging**) of the amino acid with its cognate tRNA species.
   a. The enzymes must **bind** both the **tRNA and the amino acid** with high specificity in order to properly match them.
   b. Energy from ATP hydrolysis is used to **activate the amino acid** by joining it initially with AMP to form aminoacyl-AMP.
   c. The amino acid is then transferred from aminoacyl-AMP to the **3′ acceptor arm of the tRNA.**
   d. The **aminoacyl-tRNA is "charged"** (ie, it carries the **energy** needed to form the peptide bond in its aminoacyl linkage).
2. **Proteins are costly** to the cell, requiring hydrolysis of five high-energy phosphate bonds per amino acid incorporated.
   a. Unlike fats and carbohydrates, the energy invested in protein synthesis is not recovered when the protein is degraded.
   b. This **energy** is not spent merely for synthesis of the peptide bonds, but mainly for **regulating fidelity of translation** (ie, making proteins of proper, defined sequences).

c. The **high-energy phosphates** expended to regulate fidelity are mainly contributed by hydrolysis of **GTP** through the guanosine triphosphatase (GTPase) activities of **translation factors** that control proper decoding of the mRNA, ie, base pairing of the tRNA anticodon with the codon of the mRNA.

B. **Ribosomes** are the ribonucleoprotein machines that **translate** the mRNA into polypeptide chains.
   1. The cytoplasmic ribosomes of eukaryotic cells are composed of **two subunits.**
      a. The **small (40S) subunit** is responsible for assembling the **initiation complex** with the mRNA and the initiator aminoacyl-tRNA.
      b. The **large (60S) subunit** has the **peptidyl transferase activity,** which is responsible for synthesis of the peptide bonds and is a function of its 28S rRNA.
      c. The size of the complete ribosome of the eukaryotic cytoplasm is **80S.**
   2. **Mitochondrial ribosomes** of human cells are structurally similar to those of **prokaryotes.**
      a. They are composed of **30S small and 50S large subunits** to make up a **70S complex.**
      b. Mitochondrial protein synthesis uses its own pool of rRNAs and tRNAs, many of which are actually encoded on the mitochondrial chromosomes and differ from those used by cytoplasmic ribosomes.
   3. The ribosome has **three tRNA binding sites.**
      a. The **A site or aminoacyl site** binds the incoming aminoacyl-tRNAs.
      b. The **P site or peptidyl site** binds the growing polypeptide chain still attached to the previous tRNA used, the peptidyl-tRNA.
      c. The **E site or exit site** binds the tRNA after it has disconnected from the growing polypeptide and is about to exit.

C. The first step in translation of an mRNA is assembly of an **initiation complex** of ribosome, mRNA, and initiator aminoacyl-tRNA (Figure 12–2).
   1. The **start codon** is an AUG triplet designating **methionine** that is distinguished by the **Kozak sequence,** which base pairs with a portion of the 18S rRNA of the small subunit.
   2. **Initiation factors** regulate formation of the initiation complex in a **stepwise** process.
      a. The special **initiator Met-tRNA** is first loaded into the P site of the 40S subunit.
      b. The **mRNA** then binds so that the AUG start codon is aligned with the anticodon of the initiator Met-tRNA.
      c. Once this pre-initiation complex is formed, the **large subunit** binds and forms the active **80S initiation complex.**
      d. High-energy phosphates of ATP and GTP are hydrolyzed by the **initiation factors** to ensure proper assembly of the complex.

D. Stepwise **elongation** of the polypeptide chain is a carefully regulated, **cyclic process** (Figure 12–3).
   1. The nascent chain attached to the previous tRNA used, **peptidyl tRNA,** is transferred to the **P site** of the ribosome.
   2. The codon specifying the next amino acid to be added is displayed in the **open A site.**

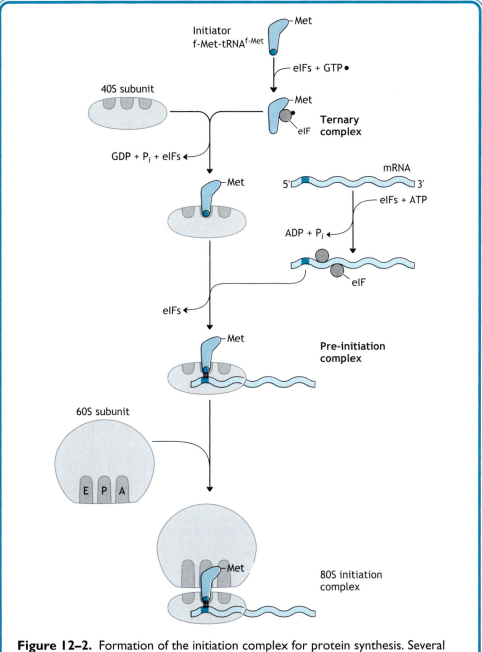

Figure 12–2. Formation of the initiation complex for protein synthesis. Several eukaryotic initiation factors (eIFs) ensure proper assembly at each step. The initiator Met-tRNA is bound in the peptidyl site of the 80S complex with its anticodon (*black stripes*) base paired to the AUG start codon (*gray box*) of the mRNA.

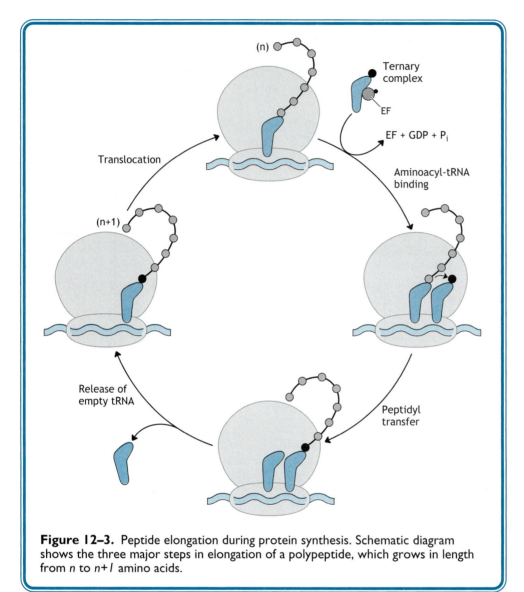

Figure 12–3. Peptide elongation during protein synthesis. Schematic diagram shows the three major steps in elongation of a polypeptide, which grows in length from *n* to *n+1* amino acids.

3. The **aminoacyl-tRNAs** are escorted to the A site by an **elongation factor** to attempt base pairing.
   a. A **good match,** as defined by proper base pairing or hydrogen bond formation **between codon and anticodon,** allows the aminoacyl-tRNA to persist in the A site long enough for **formation of the peptide bond.**
   b. A **poor codon-anticodon match** (mismatched or minimal base-pairing) causes the aminoacyl-tRNA to **dissociate** before the peptide bond can be formed.
4. Peptide bond formation **(peptidyl transfer)** occurs by reaction of the nascent peptide with the aminoacyl-tRNA properly base-paired in the A site.

5. The **ribosome then translocates** to the next codon, with the peptidyl-tRNA shifting from the A site to the P site and the now uncharged tRNA exiting the ribosome from the E site.
6. This process is repeated for all codons of the mRNA that specify amino acids, using high-energy phosphates from GTP at each cycle.

E. **Termination** of the polypeptide occurs when a **stop codon** appears in the A site and triggers release of the mRNA and completed protein.
   1. When a stop codon appears in the A site, one of several **termination factors** binds the site.
   2. This causes **reaction of the polypeptide with water,** which **releases the completed polypeptide** from the ribosome.
   3. The **ribosomal subunits dissociate** and **release the mRNA.**

INHIBITORS OF PROTEIN SYNTHESIS AS ANTIBIOTICS

- *Many of the inhibitors of protein synthesis are selective for prokaryotic ribosomes, which reduces potential for toxicity to humans.*
- ***Streptomycin, gentamicin,*** *and other aminoglycosides interfere with assembly of the 30S initiation complex and promote incorrect base pairing.*
- ***Tetracycline*** *and its derivatives inhibit entry of the aminoacyl-tRNAs into the A site of both eukaryotic and prokaryotic ribosomes, but eukaryotic plasma membranes are impermeable to these drugs.*
- ***Erythromycin*** *and the other macrolides prevent release of tRNAs from the ribosomal A site after peptide bond formation.*

III. Post-translational Modification of Proteins

A. Membrane and secretory proteins must be modified after translation to ensure proper cellular localization.
   1. **Secretory and membrane-targeted proteins** are synthesized as larger **precursors** on endoplasmic reticulum (ER)–bound ribosomes, which inject the protein across or into the ER membrane as it is translated.
   2. **Endoproteases** cleave these proteins to **activate** them in the ER during sorting in the Golgi apparatus, during storage in secretory vesicles, or at the time of use when they arrive at their final destinations.
      a. **Digestive enzymes,** such as trypsin, chymotrypsin, and ribonuclease, are made as inactive **zymogens** in the pancreas and then activated by proteolytic cleavage when they arrive in the intestine.
      b. Collagen is secreted as **procollagen** strands, which are assembled outside the cell and **trimmed** by proteases to mature collagen (see Chapter 2).

B. **Post-translational covalent modification** of many proteins is important for their **proper function** and **subcellular localization.**
   1. Most secreted and membrane-embedded proteins are modified by addition of sugar structures (**glycosylation**) to the side chains of some amino acids.
      a. Glycosylation can help **stabilize** glycoproteins against degradation or provide proper conformation for protein **function.**
      b. Oligosaccharides displayed by proteins can provide important biologic signals.
      c. A common form of protein glycosylation (**N-linked**) modifies the amide side chain of asparagine residues in many integral proteins of the plasma membrane, eg, **hormone and growth factor receptors.**

d. N-linked oligosaccharides capped with **mannose 6-phosphate** (Man-6-P) target glycoproteins for delivery to the **lysosomes.**
       (1) Lysosomal enzymes, including many hydrolases responsible for degrading cellular waste products are synthesized on ER-bound ribosomes.
       (2) These proteins are glycosylated and then the oligosaccharides are modified, including addition of Man-6-P, as they pass through the Golgi apparatus.
       (3) Man-6-P receptors bind these enzymes in the Golgi and then transport them via vesicles to lysosomes.

DISORDERS OF LYSOSOMAL ENZYME LOCALIZATION

- *Several disorders that lead to impaired lysosome function include the **mucolipidoses** I-cell disease and pseudo-Hurler polydystrophy.*
- *I-cell disease (mucolipidosis II, ML-II) is an autosomal recessive disorder in which intracellular trafficking of lysosomal enzymes is disrupted due to deficiency of one of the enzymes involved in synthesis of the Man-6-P marker.*
- *Patients with I-cell disease exhibit severe **psychomotor retardation;** coarse facial features; and skeletal malformations, including kyphoscoliosis, anterior beaking of the vertebrae, and a lumbar gibbus deformity.*
  - *Patients exhibit low birth weight and restricted growth and have a high likelihood of death by age 10.*
  - *Cells cultured from ML-II patients show **dense inclusion bodies**—hence, the term "I-cells," due to lysosomes that store, rather than degrade, cellular waste materials.*
- *Pseudo-Hurler polydystrophy (mucolipidosis III, ML-III) is related to I-cell disease, but cells of these patients retain some activity of the deficient enzyme.*
  - *Lysosome function is consequently less impaired in ML-III patients than in ML-II patients.*
  - *Symptoms show a **later onset** and **more benign clinical manifestations than in ML-II,** and some ML-III patients may reach adulthood.*
  - *In ML-III patients, stiffness of the hands and shoulders due to rheumatoid arthritis leads to **claw-hand deformities** in addition to short stature and scoliosis.*

       e. The hydroxyl groups on serine or threonine side chains can also be glycosylated (**O-linked**) on some proteins.
         (1) All **glycosaminoglycan** chains except hyaluronic acid start as O-linked glycoproteins.
         (2) The hydrophilic properties of **mucins,** the large glycoproteins of mucus, are due to multiple O-linked glycosyl chains on these proteins.
   2. Proline and lysine residues of collagen chains may be modified by **hydroxylation** (see Chapter 2).
   3. Many proteins undergo post-translational **acylation,** which is the addition of fatty acids to various amino acid side chains.
      a. Acylation has several functional effects on proteins, especially to help **anchor them to membranes.**
      b. Palmityl, myristyl, and farnesyl groups are most commonly involved in these modifications.

INHIBITORS OF FARNESYLATION AS ANTICANCER AND ANTIPARASITIC AGENTS

- *Protein farnesyltransferase (PFT) is responsible for farnesylation of cellular proteins, and PFT inhibitors have recently been developed for treatment of diseases involving farnesylated proteins.*

• *Up to 30% of cancers involve mutations of **Ras**, a small GTPase that regulates the cell cycle and cellular signaling pathways in response to growth factors (see Chapter 14).*
  – *Farnesylation of Ras mediates interaction with the cell membrane, which is required for proper Ras function.*
  – ***PFT inhibitors,** eg, tipifarnib, are effective anticancer agents, especially for acute myeloid leukemia, melanoma, and myeloma.*
• *Many **pathogenic protozoa,** including* Trypanosoma brucei *(trypanosomiasis or African sleeping sickness),* Plasmodium falciparum *(malaria),* Leishmania *species (leishmaniasis), and the intestinal parasites* Giardia lamblia *and* Entamoeba histolytica, *depend on farnesylated proteins for growth and reproduction.*
– *PFT inhibitors are effective in the treatment of many of these diseases, especially trypanosomiasis and malaria.*

    4. The carboxyl groups of specific glutamate residues in certain proteins may be **carboxylated** in a vitamin K–dependent reaction.
      a. The clotting factor **prothrombin** is carboxylated on glutamate residues, creating γ-**carboxyglutamate** groups that form binding sites for Ca^{2+} on the protein.
      b. **Calcium binding** to the bone matrix protein **osteocalcin** also depends on γ-carboxyglutamate groups.

VITAMIN K DEFICIENCY

• *Deficiency of vitamin K, which is **fat-soluble,** is rare and produces only mild symptoms such as a delay in blood clotting (**prolonged prothrombin time**) in adults.*
• *Most adults obtain all the vitamin K they need from synthesis by intestinal flora and subsequent uptake in micelles.*
• *Deficiencies characterized by **excessive bleeding** may occur in infants due to their lack of intestinal bacteria or in adults having **fat malabsorption disorders,** such as **cystic fibrosis,** which result in insufficiency of pancreatic lipase secretion.*

    5. Many proteins undergo reversible **phosphorylation** that is a major mechanism for regulation of protein function (see Chapter 14).
   C. **Protein degradation,** which regulates protein availability and, hence, gene expression in the cell, occurs by two major mechanisms.
    1. **Ubiquitin-mediated protein degradation** is responsible for regulated degradation of proteins in the **cytoplasm.**
      a. The small protein **ubiquitin** is **covalently attached** to proteins targeted for degradation eventually forming a **polyubiquitin chain** on the protein.
      b. The modified protein binds to and is taken up by the **26S proteasome,** a degradative molecular machine.
      c. Once inside the central cavity of the proteasome, the protein is degraded to its component amino acids and the ubiquitin molecules are released and reused.
    2. **Lysosomes are degradative organelles** that have a low internal pH (~5.5) and contain a battery of **hydrolases,** such as proteases, nucleases, glycosidases, and lipases.
      a. Whole organelles or vesicles fuse with the lysosomes and their contents are degraded.
      b. Proteins taken up by the lysosomes are **degraded to constituent amino acids,** many of which are released into the cytoplasm for **reutilization.**

c. Many membrane lipids, eg, ceramides, sphingomyelin and glycosphin-golipids, are also taken up into lysosomes for degradation.

GLYCOSAMINOGLYCAN ACCUMULATION DUE TO DEFECTIVE DEGRADATION IN THE MUCOPOLYSACCHARIDOSES

- *Proteoglycans **are glycoproteins that have a special type of polysaccharide called** glycosamino-glycans (GAGs), most of which are **large, unbranched polysaccharides composed mainly of a** dis-accharide repeat of a sugar acid and a hexosamine.*
- *The mucopolysaccharidoses (MPSs) are a diverse group of rare inherited diseases arising from **patho-logic accumulation of the GAGs** within the interstitium due to defective degradation arising from de-ficiencies in various lysosomal enzymes that normally degrade these GAGs within cells.*
- *As with the mucolipidoses and the enzyme-deficiency diseases (see Chapter 3), strategies using enzyme replacement therapy are being developed for treatment of many of the MPS syndromes.*
- *These disorders exhibit an overlapping array of symptoms and clinical features.*
  - *Patients appear normal at birth, but abnormalities develop either during infancy or at about 2–6 years of age.*
  - *Initial signs are dysmorphic features, especially **coarse facial features;** macrocephaly (large head); and hirsutism (excessive body hair).*
  - *Clinical manifestations may occur in virtually every organ system, with widely variable progression often involving short stature, skeletal deformities, intestinal abnormalities, spasticity, joint stiffness, and reduced life expectancy (< 20 years).*
  - ***Learning disabilities** are characteristic of some MPS disorders, and mental retardation occurs in se-vere cases.*
- ***Hurler syndrome** (MPS type I) is caused by **deficiency of α-iduronidase,** a lysosomal enzyme in-volved in degradation of dermatan sulfate and heparan sulfate, which accumulate in the cells of all tis-sues and spill over into the urine.*
  - *Distinguishing clinical features of MPS-I include corneal clouding and a particular type of acute an-gular kyphoscoliosis (combined outward and lateral spinal curvature).*
  - *MPS-I is the most common of the MPS syndromes and is also called "gargoylism" due to stooped stature and coarse facies.*
- ***Hunter syndrome** (MPS-II) is an X-linked disorder arising from **deficiency of iduronate sulfatase,** which helps degrade heparan sulfate and dermatan sulfate.*
  - *Deafness is a distinguishing feature of Hunter syndrome.*
  - *Unlike many of the MPSs, MPS-II is not associated with mental retardation.*

IV. Regulation of Gene Expression

    A. At any given time, not all the genes of an organism will be equally active in tran-scription.

        1. In prokaryotes and eukaryotes, the expression of individual genes is controlled by activation or inhibition of **RNA polymerase** on each gene by **transcription factors.**

        2. Selective regulation of gene expression in prokaryotes allows the organism to respond to changing **environmental factors,** eg, nutrient availability, by alter-ing the repertoire of proteins it makes.

            a. The **structural genes** encoding enzymes involved in a particular process tend to be located adjacent to each other so that they may be **coordinately controlled** by nearby **regulatory genes.**

    b. This organizational unit is called an **operon, eg, the *lac* or *trp* operons** that permit *Escherichia coli* to respond to availability of lactose or tryptophan.

  2. In humans, different sets of genes are turned off or on in each type of cell or tissue, both in the major process of **differentiation,** and in response to the body's more immediate **physiologic needs,** eg, growth, development, or disease.

B. The ***lac* operon** of *E coli* is a good model for regulation of prokaryotic gene expression in response to environmental cues (Figure 12–4).

  1. The *lac* operon has **three structural genes** (genes that encode protein products), the *lac*Z, *lac*Y, and *lac*A genes.

  2. The main gene, ***lac*Z, encodes β-galactosidase,** which hydrolyzes the disaccharide lactose to glucose + galactose to begin their metabolism.

  3. The **genes** of the operon become coordinately **up-regulated or induced** when **allolactose** (an **inducer** and a derivative of **lactose**) is present, indicating availability of lactose as a potential source of energy and carbon.

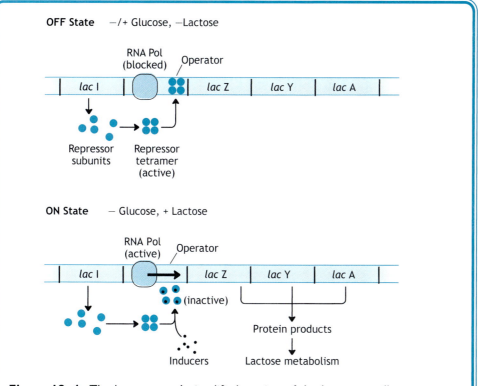

Figure 12–4. The *lac* operon. A simplified version of the *lac* operon illustrates how activity is regulated by availability of lactose as the sole carbon source. Repressor is the product of the *lac*I regulatory gene. Lactose in the environment is converted to allolactose, which acts as the inducer. The ON state can only occur in the absence of glucose. With repressor inactive (unbound), RNA polymerase can transcribe the structural genes.

a. In the absence of lactose, the gene is in the **OFF state,** with the *lac* **repressor bound to the operator** and thereby blocking transcription.

    b. Inducer binds to the repressor causing it to **dissociate from the operator** and converting the operon to the **ON state.**

    c. With the operator unblocked, **RNA polymerase** can progress beyond the promoter and **transcribe** the *lac* structural genes so that the cell is able to metabolize lactose.

C. Eukaryotic gene regulation is much more complex than in prokaryotes, with expression dependent on several types of transcription factors as well as chromatin structure.

  1. Gene expression may be controlled by **transcription factors** that bind to the **5′-untranslated region,** which encompasses the promoter.

    a. Transcription factors are mobile, capable of affecting multiple genes located on different chromosomes and are thus considered *trans*-acting elements.

    b. The sequences of the gene to which these factors bind are called *cis*-acting **elements.**

    c. Initiation of transcription of all genes is absolutely dependent on binding of multiple **general transcription factors,** which assemble into a large **preinitiation complex** that activates RNA polymerase.

    d. Expression of specific genes may also be controlled by **regulatory factors** that bind to sites distant from the promoter to turn transcription on or off in response to changing needs of the body, eg, nutrient availability, hormones, growth factors, or energy status.

  2. Enhancer regions are *cis*-acting elements, sequences that can also affect transcription.

    a. Enhancers may be located almost anywhere in the gene, including within an intron.

    b. Enhancers bind protein factors that **promote transcription** of the gene by interacting directly with RNA polymerase or with the pre-initiation complex.

  3. An example of specific transcriptional control is **cyclic AMP-dependent regulation of genes** that have a cyclic AMP **response element** (CRE) through the action of the transcription factor **CREB** (cyclic AMP response element binding protein, Figure 12–5).

D. Expression of eukaryotic genes also depends on changes in chromatin structure, called **chromatin remodeling.**

  1. Euchromatin is loosely packed, which permits access of RNA polymerase and transcription factors, and thereby promotes transcriptional activity of the genes within it.

  2. Heterochromatin is densely compacted, which limits access of transcription factors and keeps the genes within the region transcriptionally inactive.

  3. The structural differences between euchromatin and heterochromatin are coordinately regulated by reversible **covalent modification** of the DNA or histones.

    a. Methylation of DNA on certain cytosines tightens packing interactions between the protein and DNA, which inhibits genes in the region.

    b. Histone acetylation decreases the affinity of histones for the DNA and helps activate genes in the region.

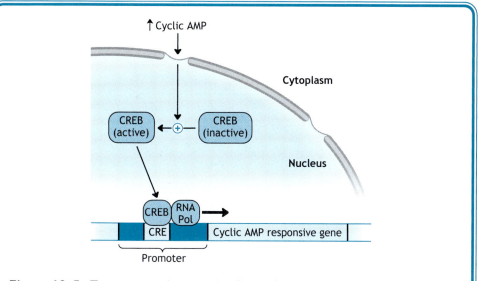

Figure 12–5. Transcriptional control by CREB. Cyclic AMP is a second messenger that mediates signaling from cell-surface receptors to elicit a response from the cell, in this case, a change in expression of genes that have a cyclic AMP response element (CRE) by binding of activated CREB.

V. Mutations

A. Single-base substitutions may occur as a result of DNA damage, chemical mutagenesis, or unrepaired errors in replication.

1. **Transitions** and **transversions** that may arise from single-base substitutions have been reviewed in Chapter 11.

2. Single-base substitutions may have no physiologic effect if they occur in a DNA region that is not part of the coding or regulatory regions of a gene.

3. Mutations may **alter regulatory sequences,** eg, in promoter or enhancer regions, which can affect gene expression.

4. Single-base changes that occur within a coding region of a gene *may* produce **disease alleles** (Figure 12–6).

   a. **Nonsense mutations** change a codon specifying an amino acid to a stop codon, which terminates translation and causes production of a truncated protein.

   (1) **Truncated proteins** may have decreased activity or be hyperactive relative to the full-length protein, and adverse effects on the cell may occur either way.

   (2) Several variants of **β-thalassemia** are caused by nonsense mutations leading to production of truncated, unstable β-globin chains.

   b. **Missense mutations** change the codon specificity from one amino acid to another, which alters the protein sequence and may also affect its function, eg, in sickle cell anemia.

Normal	AUG-UCA-AGA-CCG-AAC-GGC-UAC-UUC-GAU-CUA-AUA
	Met-Ser-Arg-Pro-Asn-Gly-Tyr-Phe-Asp-Leu-Ile

Single-base substitutions

Silent	AUG-UCA-AGA-CCG-AAU-GGC-UAC-UUC-GAU-CUA-AUA
	Met-Ser-Arg-Pro-Asn-Gly-Tyr-Phe-Asp-Leu-Ile
Missense	AUG-UCA-AGA-CCG-AAC-GUC-UAC-UUC-GAU-CUA-AUA
	Met-Ser-Arg-Pro-Asn-Val-Tyr-Phe-Asp-Leu-Ile
Nonsense	AUG-UCA-AGA-CCG-AAC-GGC-UAG-UUC-GAU-CUA-AUA
	Met-Ser-Arg-Pro-Asn-Gly-STOP
	Truncated protein

Insertions and deletions

Insert (+1 nt)	AUG-UGC-AAG-ACC-GAA-CGG-CUA-CUU-CGA-UCU-AAU-A
	Met-Cys-Lys-Thr-Glu-Arg-Leu-Leu-Arg-Ser-Asn-
	Garbled protein sequence
Delete (−1 nt)	AUG-UCA-AGA-CCGA-ACG-GCU-ACU-UCG-AUC-UAA-UA
	Met-Ser-Arg--Arg-Thr-Ala-Thr-Ser-Ile-STOP
	Garbled sequence, truncated protein

Splicing error

	AUG-UCA-AGA-CCG-AAC-GGC-AGA-AGC-UGU-GUC-AAA
	Met-Ser-Arg-Pro-Asn-Gly-Arg-Ser-Cys-Val-Lys
	Garbled sequence

Figure 12–6. Mutations of gene sequences that may affect protein function and cause disease.

    c. Silent mutations, which often occur in the 3′ base of a codon, do not alter codon specificity, so there is no effect on the protein's sequence.

B. Insertions and deletions may cause production of altered proteins that have either subtle or drastic functional changes.

    1. If the DNA sequence is altered by deletion or insertion of 3*n* nucleotides, then the mutant protein will lack or gain *n* amino acid(s), but its sequence will otherwise be normal.

    2. Insertion or deletion of a number of nucleotides that is not divisible by three will cause a **frameshift** such that different, **garbled protein** sequence will be synthesized downstream of the mutation.

C. **Splicing errors** alter the critical sequence around an intron-exon splice junction.
  1. This may be caused by single-base substitutions, insertions, or deletions.
  2. Creation of an abnormal splicing site or destruction of the normal site may result in **incorporation of an intron** into a "finished" mRNA.
  3. Translation of the intron region of the mutant mRNA produces a **garbled protein sequence** until an **in-frame stop** codon causes **termination** of the truncated, mutant protein.

CLINICAL PROBLEMS

A 9-year-old boy is referred for evaluation of his hearing. A note from his school principal explains that he is inattentive in class. Initial physical examination indicates that he is at the 10th percentile for height, has coarse facial features, and is somewhat macrocephalic; however, the remainder of the examination is within normal limits. Audiometry results confirm partial bilateral deafness, which is sensorineural in etiology. An IQ examination shows that he is in the 60th percentile for intelligence. Family history of mucopolysaccharidoses prompts specialty testing, which indicates elevated levels of dermatan sulfate and heparan sulfate in both a skin biopsy and urine sample.

1. Biochemical analysis of a skin biopsy from this patient would most likely indicate a deficiency of which of the following enzymes?
   A. β-Galactosidase
   B. α-L-Iduronidase
   C. Iduronate sulfatase
   D. N-Acetylgalactosamine sulfatase
   E. β-Glucuronidase

The sickled shape of erythrocytes in patients with sickle cell anemia occurs because of the tendency for HbS to polymerize. HbS differs from HbA by substitution of a solvent-exposed glutamate by valine in β-globin, which forms a "sticky" patch that promotes aggregation and polymerization of the protein.

2. The genetic change that produced the mutant hemoglobin in sickle cell anemia can be classified as which type of mutation?
   A. Silent
   B. Missense
   C. Nonsense
   D. Insertion
   E. Deletion

Infections by the ulcer-causing bacterium *Helicobacter pylori* can be treated effectively with a prolonged course of doxycycline or another of the tetracycline family of antibiotics, potent inhibitors of prokaryotic protein synthesis.

3. Which of the following explains why tetracycline is selective for prokaryotes and minimally toxic to humans?

   A. It is ineffective against the 70S ribosomes.

   B. It is ineffective against the mitochondrial ribosomes.

   C. It only inhibits prokaryotic peptidyl transferase.

   D. It cannot pass across eukaryotic membranes.

   E. It blocks the A site only of prokaryotic ribosomes.

Some patients with familial hypercholesterolemia produce a truncated form of the LDL receptor, termed the "Lebanese" allele, which lacks three of the five domains of the protein and causes it to be retained in the endoplasmic reticulum. Analysis of the mutant gene indicated that the sequence of the protein was normal up to the point where it terminated.

4. The genetic change that produced the mutant LDL receptor in these cases can be classified as which type of mutation?

   A. Silent

   B. Missense

   C. Nonsense

   D. Insertion

   E. Deletion

A 2-year-old boy in whom Down syndrome was diagnosed when he was an infant comes in for a check-up. Although he is developmentally delayed indicating potential mental retardation, he is exhibiting some clinical features that are inconsistent with Down syndrome. These features include coarse facial features, small stature, radiographic evidence of kyphoscoliosis, widening of the ribs, and malformed vertebrae.

5. Microscopic examination of skin or muscle biopsy specimens from this patient would be likely to reveal dense inclusions corresponding with which organelles?

   A. Mitochondria

   B. Nuclei

   C. Golgi apparatus

   D. Lysosomes

   E. Peroxisomes

A 17-year-old woman with cystic fibrosis is evaluated for knee pain. On review of systems, she also notes persistent bleeding from cuts in her skin and bleeding of her gums after brushing her teeth. Physical examination is remarkable for an obviously swollen right knee that is tender with limited range of motion. Fluid drained from the knee is bloody (hemarthrosis). Her complete blood count is normal, but prothrombin time is elevated.

6. This patient appears to be suffering from a deficiency of which of the following vitamins?

   A. Vitamin A

   B. Vitamin B_{12}

   C. Vitamin C

D. Vitamin D

E. Vitamin K

ANSWERS

1. The answer is C. This patient's clinical presentation is consistent with one of the mucopolysaccharidoses, but it can be difficult to determine which type given the wide variability of expression of these disorders. One clue is provided by the hearing loss, a characteristic feature of MPS-II, Hunter syndrome. In addition, his above-average intelligence for his age group and the absence of scoliosis distinguish MPS-II from MPS I, the Hurler-Scheie syndromes. The latter are characterized by mental retardation to varying degrees. The patient appears to have a severe form of Hunter syndrome, so the cells of his tissues should be deficient in the lysosomal enzyme iduronate sulfatase.

2. The answer is B. A missense mutation results from a change in codon specificity from one amino acid to another. This alters the protein sequence and may affect the protein's structure and function. By definition, substitution of valine for glutamic acid in the β-globin molecule represents a missense mutation at the level of the gene. Sickle cell anemia illustrates how important even a single amino acid in a large protein can be to the function of the protein and the physiology of the cell. However, it is more common to find that missense mutations have less dramatic effects than in this case.

3. The answer is D. Tetracycline antibiotics operate by blocking the aminoacyl binding site of 30S ribosomes found both in prokayotes and in the mitochondria of eukaryotes. However, the drug may be used as a selective antibiotic with minimal toxicity to patients because it cannot pass through the plasma membranes of human cells. If it could do so, the drug would be cytotoxic because it would interfere with mitochondrial function by inhibiting protein synthesis on the 70S ribosomes of the organelles.

4. The answer is C. Production of a truncated protein indicates that a mutation has occurred, but this phenomenon may have arisen from a frameshift mutation (insertion or deletion) or by a nonsense mutation. The most likely possibility is a nonsense mutation because sequence analysis of the truncated protein showed that it had normal (wildtype) sequence. Insertion and deletion events often produce a stretch of garbled or abnormal protein sequence at the C-terminal end of the truncated protein arising from out-of-frame translation of the mRNA downstream of the mutation until a stop codon is encountered.

5. The answer is D. As this patient ages, a variety of skeletal defects and short stature that are consistent with a lysosomal storage disease (mucolipidosis), either I-cell disease or pseudo-Hurler polydystrophy, are developing. Both diseases arise from a deficiency of an enzyme involved in synthesis of the Man-6-P marker on lysosomal enzymes. Such "misaddressed" proteins are secreted rather than trafficked to the lysosomes. The degradative function of lysosomes is impaired as a result and the organelles tend to accumulate waste products (hence, the term "storage disease"). It is these inclusion bodies or dense structures that would be visible by microscopic examination of the patient's cells in a biopsy specimen.

6. The answer is E. The patient's symptoms and prolonged prothrombin time suggest that she has a mild coagulation disorder potentially due to vitamin K deficiency. Several coagulation factors including prothrombin require carboxylation on glutamic acid residues for optimal function. These proteins are carboxylated in vitamin K–dependent reactions. Vitamin K deficiency may occur in people suffering from cystic fibrosis, which causes gastrointestinal complications due to pancreatic insufficiency. Secretion of pancreatic enzymes such as lipase, which releases fatty acids from triglycerides to facilitate absorption from the gut, is impaired in cystic fibrosis patients. This fat malabsorption condition has manifested itself in this patient's case as a deficiency in the fat-soluble vitamin K. Although bleeding gums are one characteristic of scurvy, other manifestations of vitamin C deficiency, eg, loose teeth, are absent in this case.

CHAPTER 13
HUMAN GENETICS

I. Overview of Mendelian Inheritance

A. A **gene** is defined as a unit of DNA that encodes an RNA product.
  1. The RNA product may encode transfer RNAs (tRNAs), ribosomal RNAs (rRNAs), or small nuclear RNAs (snRNAs) that have end point functions in the cell.
  2. If the RNA product is a messenger RNA (mRNA), then it must be translated into a protein to complete expression of the gene.
  3. Variants of a gene that differ in DNA sequence among individuals in the population are called **alleles.**
     a. The single most prevalent version of the gene in the population is referred to as the **wild-type ("normal")** allele.
     b. If there is more than one common version of the gene in the population, these are called **polymorphisms.**
     c. Mutant alleles are versions of the gene that differ in sequence from the wild-type allele and that produce **defective** products.
     d. The chromosomal location of a gene is its **locus.**

B. The set of alleles that make up the genetic composition of a person is called the **genotype,** which may refer either to all genes or to a specific gene or locus.
  1. The diploid content of human cells is 46 chromosomes—22 autosomal pairs and 2 sex chromosomes (XX in females, XY in males).
  2. For genes located on the **autosomes,** the genotype at a locus is formed from two alleles.
  3. Each parent contributes one allele through **random segregation** of chromosomes during meiosis.
  4. If both alleles at a locus are identical, the person is said to be **homozygous** for that gene.
  5. In the case where the two alleles are different, the person is **heterozygous** for that gene.
  6. Since males have only one X chromosome, they usually have only a single allele and are thus **hemizygous** for all X-linked genes.

C. The measurable expression of the genotype as a molecular, clinical, or biochemical **trait** is the **phenotype.**

D. **Pedigree analysis** evaluates transmission of a single-gene disorder within a family or **kindred** (Figure 13–1).

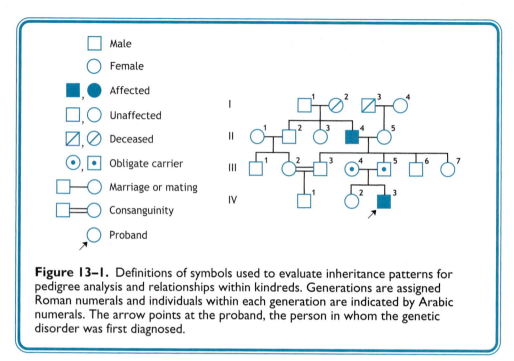

Figure 13–1. Definitions of symbols used to evaluate inheritance patterns for pedigree analysis and relationships within kindreds. Generations are assigned Roman numerals and individuals within each generation are indicated by Arabic numerals. The arrow points at the proband, the person in whom the genetic disorder was first diagnosed.

II. Modes of Inheritance in Single-Gene Disorders

A. In **autosomal recessive** inheritance, the condition is expressed only in persons who have two copies of (ie, are **homozygous** for) the mutant allele (Figure 13–2).

   1. Autosomal recessive inheritance is often observed with **enzyme deficiencies,** where heterozygotes express 50% of normal activity.

   a. However, **50% of normal enzyme activity** in these cases permits normal physiologic function because expression of enzyme from the normal allele is sufficient to provide for the needs of the cell.

   b. This phenomenon is often called the **margin of safety** effect.

   2. Both parents of an affected person for an autosomal recessive disorder must have one normal and one mutant allele, making them **obligate carriers** barring very rare new mutations.

   3. The likelihood of a person being homozygous for an autosomal recessive trait increases in **consanguineous matings** because of the existence of a **common ancestor.**

   4. Rare autosomal recessive diseases also occur with high frequency among **genetically isolated populations** due to **inbreeding.**

TAYS-SACHS DISEASE IN A GENETICALLY ISOLATED POPULATION

• The biochemical defect in Tay-Sachs disease is an inherited deficiency of **β-hexosaminidase,** a lysosomal enzyme responsible for hydrolysis of GM_2 ganglioside, which accumulates abnormally in the lysosomes.

• Children with Tay-Sachs disease exhibit hypotonia (poor muscle tone) and progressive **neurologic symptoms,** including blindness and **mental retardation.**

CLINICAL CORRELATION

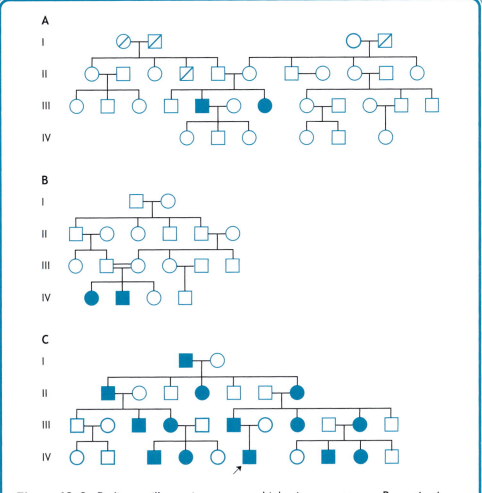

Figure 13–2. Pedigrees illustrating autosomal inheritance patterns. Recessive inheritance is shown in pedigrees A and B. Note that consanguinity in pedigree B reinforces the hypothesis of an autosomal recessive disorder. Dominant inheritance is shown in pedigree C, in which every affected person has an affected parent.

- *Most patients are diagnosed at 5–6 months and do not live beyond their second year.*
- *This autosomal recessive disease occurs in **Ashkenazi Jews, the Pennsylvania Amish,** and several other populations with an incidence of 1 in 3600, 100 times higher than the overall population; the carrier frequency in these populations is about 3%.*

 5. Pedigree charts for an autosomal recessive disorder may show the following:
  a. The disease phenotype is expressed by siblings but not by their parents or offspring.
  b. Equal occurrence in males and females.
  c. Recurrence risk for each sibling is 25%.
  d. Possible consanguinity.

B. In **autosomal dominant** inheritance, the condition is expressed even if a single mutant allele is present, ie, in the **heterozygous state** (Figure 13–2).

1. Following are at least four possible situations by which having one normal copy of a gene is insufficient to prevent disease, leading to a dominant phenotype (Table 13–1):

   a. When the presence of 50% normal activity (ordinarily the margin of safety) is not generous enough to allow normal physiologic function, a condition called **haploinsufficiency.**

   b. When the defective allele produces a **malfunctioning protein product** that binds to and **interferes** with function of the normal gene product—the **dominant negative effect.**

   c. When the mutant protein has an enhanced function that overrides normal controls or is **cytotoxic.**

   d. When the phenotype appears as dominant inheritance even though the actual allele is recessive at the level of function in individual cells.

2. The **homozygous mutant state** usually produces a **more severe clinical condition** than the heterozygous condition in autosomal dominant diseases.

3. Pedigree charts for an autosomal dominant disorder may show the following features:

   a. The disease phenotype appears in **all generations,** with each affected person having an affected parent.

   b. There is an equal occurrence in males and females, except in cases when expression of the trait is influenced by the person's sex (ie, **sex-limited**).

   c. Risk of transmission of the mutant allele is 50%, but because there usually are so few persons in a family, there may be deviations from this expectation.

   d. Potential for some cases to be due to a **new mutation,** which is more likely for a dominant condition because disease symptoms would be expressed in heterozygotes.

Table 13–1. Molecular phenotypes of autosomal dominant disease.[a]

Molecular Explanation	Disease and Gene
Haploinsufficiency, or when 50% of normal gene activity is inadequate	α-Thalassemia trait and the α-globin gene β-Thalassemia trait and the β-globin gene
Dominant negative effect, when the mutant protein interferes with function of the normal protein	Osteogenesis imperfecta and the collagen IA gene (COLIAI) Marfan syndrome and the fibrillin-1 gene (FBN1)
Cytotoxic effect due to a dysregulated, mutant protein	Huntington disease and the huntingtin gene (HD)
Dominant effect at the cellular level of a recessively inherited loss-of-function mutant of a tumor suppressor gene (see Chapter 14)	Retinoblastoma and RB1 Li-Fraumeni syndrome and TP53

[a]These genetic diseases are examples of the various molecular explanations for dominant inheritance.

FIBRILLIN DEFECTS IN MARFAN SYNDROME

- *Marfan syndrome is a **connective tissue disorder** with manifestations in many organs, but especially the skeleton, blood vessels, eyes, and lungs.*
- *Many tissues, such as **lung, blood vessels,** and **skin, require elasticity** for proper function; this property is fulfilled by the matrix **elastic fibers,** which are composed of the proteins **elastin** and **fibrillin.***
  - *Marfan syndrome arises from a mutation in the gene encoding **fibrillin-1** (FBN1).*
  - *The pattern of inheritance of Marfan syndrome is **autosomal dominant** due to the failure of elastic fibers to assemble properly upon interaction of mutant fibrillin with normal elastin.*
- *The disease is usually diagnosed by adolescence, and patients exhibit tall stature and a variety of **skeletal deformities,** including very long, thin bones of the digits and limbs; flat feet; scoliosis; and breastbone deformation.*
  - ***Joint hypermobility** and a positive wrist/thumb sign are evident.*
  - *The upper segment is the distance from the top of the head to the top of the pubic symphysis; the lower segment is the distance from the top of the pubic symphysis to the floor. The **upper segment to lower segment ratio** in persons with Marfan syndrome is low (< 0.9) because the arms and legs are long relative to the torso.*
- *Characteristic ocular features of Marfan syndrome, such as **ectopia lentis** (upward lens dislocation instead of downward dislocation as in homocystinuria) and **myopia,** arise from the effects of defective fibrillin in the elastic fibers of the lens.*
- *The major cardiovascular manifestations are **mitral valve prolapse** and loss of elasticity of the aortic root, which may lead to progressive aneurysm and potentially fatal **aortic dissection.***

    C. Most **X-linked** diseases show a recessive inheritance pattern (Figure 13–3).

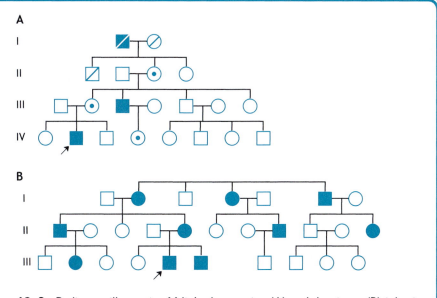

Figure 13–3. Pedigrees illustrating X-linked recessive (A) and dominant (B) inheritance patterns. Note the absence of male-to-male transmission in both pedigrees and the predominance of affected males over females in the X-linked recessive pedigree.

1. A distinguishing feature of these diseases is that there can be **no male-to-male transmission** because the sex of male offspring is determined by contribution of a Y chromosome from the father.

      2. Because they have only one X chromosome, the sons of heterozygous mothers have a 50% chance of being affected.

      3. Pedigree charts for an X-linked recessive disorder may show the following features:

        a. Incidence of disease is higher in males than in females.

        b. Female heterozygotes are usually unaffected carriers.

        c. Affected men transmit the gene to all daughters, but never to sons.

        d. New mutations cause a significant number of isolated cases in males due to unopposed expression of the mutant allele.

  D. X-linked dominant diseases are relatively **rare** (Figure 13–3).

      1. Such genes may be transmitted either to sons or daughters by an affected mother but only to daughters by an affected father.

        a. The mechanisms at the molecular and cellular levels that produce a dominant phenotype are the same as in autosomal dominant disorders.

        b. Only a few such disorders are known, including the **Xg blood group** and **vitamin D–resistant rickets.**

      2. Pedigree charts for an X-linked dominant disorder may show the following features:

        a. All daughters of affected men are affected but never their sons, which may lead to prevalence of affected females over affected males.

        b. Recurrence risk is 50% for both male and female offspring of an affected female.

        c. Absence of affected males in several generations may suggest prenatal lethality for the hemizygous state.

  E. Incompletely dominant disorders occur in cases where the heterozygous genotype produces a **different phenotype** from that seen in the homozygous genotype.

      1. The effect is often of **intermediate severity** between the unaffected and fully affected phenotypes.

      2. For example, in **sickle cell anemia,** the normal allele is incompletely dominant in heterozygotes.

        a. At the molecular level, the presence of some abnormal hemoglobin distorts some RBCs.

        b. This causes some sickling and the mild anemia characteristic of the **sickle trait.**

  F. Mitochondrial disorders exhibit a maternal inheritance pattern.

      1. Mitochondria each have at least one and often several chromosomes that have genes important for function of the organelle.

        a. The mitochondrial chromosome (**mtDNA**) is a 16.5 kb circular **plasmid.**

        b. The mtDNA bears **37 genes** encoding rRNAs, tRNAs, and some genes for proteins involved in oxidative phosphorylation.

      2. Mitochondrial disorders are **maternally transmitted** because the **ovum provides all mitochondria to the fertilized embryo** (Figure 13–4).

      3. In these disorders, affected cells usually have a mixture of mitochondria, some with mutant mtDNA and others with wild-type mtDNA, a condition called **heteroplasmy.**

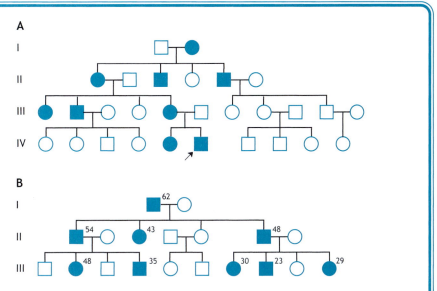

Figure 13–4. Pedigrees illustrating inheritance of (A) a mitochondrial disorder and (B) an autosomal dominant disorder exhibiting anticipation. In pedigree A, note the similarity to the X-linked dominant inheritance pattern (Figure 13-3A), but incomplete penetrance as exemplified by individuals II-4 and III-4. In pedigree B, the age of onset, indicated next to the symbols for affected individuals, becomes progressively earlier with each generation.

    a. Segregation of mitochondria during cell division is not as tightly controlled as for nuclear chromosomes, leading to random distribution of mitochondria carrying normal and mutant mtDNA to ova.

    b. This contributes to **variable expression** and **reduced penetrance** of the phenotype among persons within kindreds with mitochondrial disorders.

MITOCHONDRIAL MYOPATHY AND NEUROPATHY

- *Mitochondrial diseases are caused by mutations in various mtDNA-encoded genes, most of which result in defective mitochondrial protein synthesis.*
- *The pathology is due to decreased mitochondrial function, eg, **impaired oxidative phosphorylation**, and thus manifests in energy-intensive tissues, such as **muscles** and **nerves.***
- *Microscopic examination of a muscle biopsy specimen may show ragged red fibers (distorted, dysfunctional mitochondria).*
- *Mitochondrial diseases may be manifest as ready fatigability; elevated lactic acid levels in blood; increased muscle enzymes in serum; ataxia; and a variety of neurosensory deficits, including blindness and deafness.*
  - ***MERRF** (myoclonic epilepsy with ragged red fibers) is characterized by weakness on exertion, ataxia, and associated deafness and is due to mutation of the mitochondrial tRNA^{Lys} gene.*
  - ***MELAS** (mitochondrial encephalomyopathy with lactic acidosis and stroke-like episodes) results from a point mutation in the mitochondrial tRNA^{Leu} gene.*

– **LHON** *(Leber's hereditary optic neuropathy) causes blindness arising from mutations in the* ND1 *gene encoding complex I of the electron transport chain (see Chapter 7).*

III. Major Concepts in Human Genetics

A. When **similar phenotypes** or disease conditions can be caused by **different genotypes,** this may produce **heterogeneity.**

1. **Allelic heterogeneity** occurs when different alleles of the same gene produce clinically similar conditions.

   a. Many patients categorized as having autosomal recessive disorders actually have two different mutant alleles of the disease gene and are therefore **compound heterozygotes.**

   b. Allelic heterogeneity may account for phenotypic variability in some families with genetic disease.

2. **Locus heterogeneity** refers to the condition when mutations of more than one gene or locus can produce similar disease states.

   a. This genotypic variability is responsible for different inheritance patterns of some disorders.

   b. For example, **Ehlers-Danlos syndrome** (see Chapter 2) may be caused by mutations at more than 10 known loci, producing inheritance patterns ranging from autosomal recessive or dominant to X-linked.

B. **Variable expression** arises when the nature and severity of the phenotype for a genetic condition varies from one person to another.

VARIABLE EXPRESSION IN NEUROFIBROMATOSIS TYPE I

CLINICAL CORRELATION

- *The biochemical defect in type I neurofibromatosis (NF) involves loss-of-function mutations of the* **NF1 tumor suppressor gene.**
- *Approximately 50% of cases involve new mutations and even in cases of inherited type I NF, variability of expression in kindreds makes genetic counseling very difficult.*
- *Type I NF is an autosomal dominant disorder characterized by a wide range of clinical presentations by patients.*
  - *Among the symptoms seen in patients with type I NF are hyperpigmented skin lesions (**café-au-lait spots**), benign skin tumors (**neurofibromas**), **dysplasia** of the sphenoid bone, and benign tumors of the iris (hamartomas or **Lisch nodules**).*
  - *Patients also may suffer from mental retardation, but the main concern is a high risk of potentially fatal CNS tumors.*

C. **Pleiotropy** refers to a condition in which a mutant allele may have different phenotypic effects in various organ systems in an affected person.

D. **Genomic imprinting** is a complex phenomenon by which expression of an allele differs depending on whether it is inherited from the mother or the father.

1. A gene that is shut off when inherited from the mother is **maternally imprinted.**

2. A gene that is silenced when inherited from the father is **paternally imprinted.**

3. Imprinting involves an **epigenetic** mechanism, ie, an alteration in phenotype that does not result from a change in the genotype.

   a. Expression of the **imprinted genes** is **silenced** or shut off by **methylation** of certain chromatin regions after DNA replication during gametogenesis.

   b. The imprint is **reversible** upon passage through gametogenesis in the next generation.

DISORDERS THAT EXHIBIT IMPRINTING

- *Two clinically distinguishable conditions arise from deletion of the same region of chromosome 15 (15q11–q13) depending on whether the defective chromosome is inherited maternally or paternally, which indicates an imprinting phenomenon for one or more genes in the region.*
- ***Prader-Willi syndrome*** *arises when deleted chromosome 15 is paternally inherited.*
  - *The precise biochemical defect for Prader-Willi syndrome is unknown, but this region of chromosome 15 encompasses an imprinting control center and at least 6 maternally imprinted genes.*
  - *Patients with Prader-Willi syndrome exhibit failure to thrive and short stature initially, which converts to a tendency toward excessive eating, obesity, mild-to-moderate mental retardation, hypogonadism, and characteristic facial dysmorphology.*
- ***Angelman syndrome*** *occurs when chromosome 15 with the deletion is maternally inherited.*
  - *This disorder appears to involve paternal imprinting of UBE3A, encoding a ubiquitin-protein ligase so that this gene product must be produced from the maternal chromosome.*
  - *Angelman syndrome is a devastating neurologic disorder featuring severe mental retardation, a "happy puppet" demeanor, seizures, ataxic gait, and aphasia.*

    E. Certain inherited disorders exhibit increased severity of phenotype or decreased age of onset as the disease gene is passed from one generation to the next, an effect known as **anticipation** (Figure 13–4).

        1. Examples of genetic diseases that show anticipation are **Huntington disease, Fragile X syndrome,** and other disorders that arise from **trinucleotide repeat expansion** (see Chapter 11).

        2. Disease symptoms occur only when the length of the trinucleotide repeat region exceeds a threshold.

    F. **Mosaicism** is defined as the presence of cells in the body that are genetically different.

        1. In **somatic mosaicism,** mutation of a gene occurs in a non-germline (somatic) cell at some point during **early development** of the person, and all cells descendent from that progenitor are genetically distinct.

        2. In **germline mosaicism,** a mutation that occurred in the parent's **gonadal** make-up is transmitted through some gametes, but not all.

        3. **All females are technically mosaic** for the genes of their X chromosomes due to **inactivation** of one or the other X chromosomes early in development, a phenomenon termed the **Lyon hypothesis** or **lyonization.**

            a. Because X inactivation is random, this phenomenon accounts for **variable expression** of some **X-linked disorders,** depending on whether the disease allele or wild-type allele was inactivated.

            b. Distribution of cells from the early embryo to the tissues may be imbalanced, so that expression of the disease phenotype is not uniform among the organs.

        4. Up to 25% of patients with **Turner syndrome** exhibit a **mosaic karyotype,** in which only some cells have the 45,X karyotype classically associated with the condition.

    G. **Uniparental disomy** refers to a condition in which one or more cells of the body have two **identical chromosomes** derived from a **single parent,** which increases the likelihood of expression of recessive alleles inherited from that parent.

UNIPARENTAL DISOMY IN BECKWITH-WIEDEMANN SYNDROME

- *Children afflicted with Beckwith-Wiedemann syndrome (BWS) show an **overgrowth** condition from birth as well as **macroglossia** (enlarged tongue).*

- BWS is associated with severe hypoglycemia that may become life-threatening in addition to an enhanced tendency to develop **cancers of the liver, kidney,** and **adrenal glands.**
- The gene for BWS has been mapped to chromosome 11 (11p15), a region encompassing the gene for insulin-like growth factor II (IGF2), which is **maternally imprinted** and thus is expressed only when paternally inherited.
- Uniparental disomy may contribute to BWS in that excess paternal or decreased maternal contributions of chromosome 11 have been observed in some patients.

IV. Population Genetics: The Hardy-Weinberg Law

A. The Hardy-Weinberg Law allows calculation of genotypes based on the **allele frequencies in a population** for a given genetic disorder.

   1. If there are two alternative alleles for a gene, A and a, in a population, then we can define p as the frequency of allele A and q as the frequency of allele a, where $p + q = 1$.

   2. The following **binomial expression** governs the frequencies of the various genotypes in the population.

$$(p + q)^2 = p^2 + 2\,pq + q^2$$

   3. Thus, the chances of occurrence of AA and aa homozygotes are p^2 and q^2, respectively, whereas the chances of finding an Aa heterozygote are $2pq$.

B. The Hardy-Weinberg Law is especially useful during **genetic counseling for autosomal recessive disorders,** in which heterozygotes cannot be distinguished in phenotype from normal homozygotes (Figure 13–5).

   1. For example, in the United States, the incidence of live births of babies of Northern European descent who have cystic fibrosis (q^2, for the aa genotype) can be estimated as approximately 1 in 2500 (0.04%).

   2. The Hardy-Weinberg Law can then be used to calculate the gene frequencies for a and A (q and p):

$$q = \sqrt{\frac{1}{2500}} = \sqrt{0.0004} = 0.02, \text{ and: } p = 1-q = 1-0.02 = 0.98$$

   3. The frequency of heterozygotes having the pq genotype (unaffected carriers) can be estimated as ~4% using Hardy-Weinberg principles based on these same data.

$$2pq = 2(0.02)(0.98) = 0.0392 \text{ or } 3.9\%$$

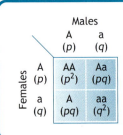

Figure 13–5. Punnett square showing application of the Hardy-Weinberg Law. The allele frequencies p and q are assumed to be equal for males and females within the population.

C. The Hardy-Weinberg Law is based on several important assumptions about the population and mating dynamics within it.

1. Matings are assumed to be **random** among persons within the population.

2. Allele frequencies are assumed to be relatively **constant** for the gene in question because of the following:

   a. The mutational rate at the locus is minimal.

   b. Persons with all genotypes are capable of passing on their alleles.

   c. The immigration rate for the population is low.

D. The **Hardy-Weinberg equilibrium** for particular alleles within a population may be **disturbed** by several factors that violate the following assumptions:

1. Stratified or isolated populations that tend to mate within the group.

2. Assortative mating, by which persons tend to choose mates who resemble themselves.

3. Consanguinity.

4. Preservation of **mutations** as alleles in the population.

5. Selection for or against certain genotypes.

   a. Dominant mutations may be subject to **negative selection** because they are expressed in the heterozygous state.

   b. Selection against homozygous affected persons in an **autosomal recessive disorder** usually has little effect on Hardy-Weinberg equilibrium because so few of the disease alleles are found in the homozygous state.

   c. Several diseases exhibit **positive selection** for the heterozygous state (**heterozygote advantage**), including sickle cell anemia and the thalassemias.

CLINICAL PROBLEMS

A couple undergoes genetic counseling for consultation on the possibility that they may have a child with Friedrich's ataxia, an autosomal recessive neuromuscular disorder. The disorder has been diagnosed in the wife's sister, and they are concerned that she is a carrier. The incidence of the disorder in the population is 1 in 25,000 live births.

1. What is the probability that this couple's first child will have Friedrich's ataxia? [Assume that the couple is not consanguineous].

   A. 1/300

   B. 1/500

   C. 1/750

   D. 1/3000

   E. 1/6000

A 6-year-old boy shows signs of significant developmental delay. After interviewing the boy's parents, a family history of varying degrees of cognitive impairment becomes apparent; specifically, the patient's father, maternal grandfather, aunt, uncle, and several cousins are affected.

2. Based on these findings, what is the most likely mode of inheritance for the disorder?

A. Autosomal recessive

B. Autosomal dominant

C. X-linked recessive

D. X-linked dominant

E. Mitochondrial

3. The incidence of sickle cell anemia among blacks is 1 in 400. Calculate the frequency of heterozygotes for sickle cell gene, encoding β_S-globin, in this population.

A. 1 in 6

B. 1 in 11

C. 1 in 20

D. 1 in 22

E. 1 in 44

The following pedigree shows the pattern of inheritance of a neurologic disorder in a large Central American kindred. Note that the numbers above symbols representing affected persons in this pedigree indicate age of diagnosis.

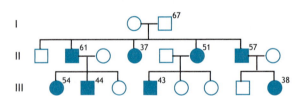

4. What type of molecular phenomenon is most likely responsible for this disorder?

A. Trinucleotide repeat expansion

B. Parental imprinting

C. Reduced penetrance

D. Multifactorial inheritance

E. Haploinsufficiency

Several members of a family have type I neurofibromatosis in which DNA analysis indicates that a son and daughter have inherited the disease from their father. The 16-year-old daughter has a pelvic neurofibroma, her brother exhibits only café-au-lait spots and a hamartoma, while the father has no detectable symptoms.

5. Type I neurofibromatosis in this family exhibits which of the following phenomena?

A. Mosaicism

B. Anticipation

C. Silencing

D. Variable expression

E. Uniparental disomy

You have assembled a pedigree for the occurrence of hemophilia A in a family of Greek-American heritage.

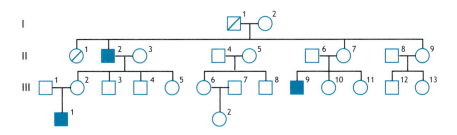

6. Based on these data, what is the most likely mode of inheritance for this disorder?
   A. Autosomal recessive
   B. Autosomal dominant
   C. X-linked recessive
   D. X-linked dominant
   E. Mitochondrial

7. What is the genotype at the hemophilia A locus of individual II-7 in the pedigree shown above?
   A. Homozygous—AA
   B. Homozygous—aa
   C. Hemizygous—Ao
   D. Hemizygous—ao
   E. Heterozygous—Aa

A 5-year-old boy complains of "being tired all the time." His mother insists that he eats well but that he sleeps 11–14 hours per day and "can't keep up with his friends on the ball fields." Physical examination reveals that he is at the 30th percentile for height and weight, with mild bilateral deafness and ataxic gait. After interviewing his mother, it becomes apparent that several other members of this family were deaf by their mid-twenties and "a muscle problem" was diagnosed in an uncle.

8. A pedigree for this family would likely reveal which of the following features?
   A. Predominant occurrence in females
   B. Absence of male-to-male transmission
   C. Lack of expression in heterozygous females
   D. Occurrence in siblings but not in their parents
   E. Tendency for the disease to skip generations

9. Angelman and Prader-Willi syndromes are related disorders affecting genes on chromosome 15 by which of the following epigenetic mechanisms?

A. Mosaicism

B. Histone acetylation

C. Haploinsufficiency

D. Imprinting

E. Viral infection

ANSWERS

1. The answer is D. The probability that the wife has inherited the disease gene from one of her parents and is a carrier is 2/3. This couple is not consanguineous, so the likelihood that he carries the disease allele is equivalent to the gene frequency in the population. The probability that either of them would pass the allele to their offspring is 1/2. Thus, the calculated overall probability of having an affected child is:

$$P = \left(\frac{1}{2}\right)\left(\frac{1}{2}\right)\left(\frac{2}{3}\right)\sqrt{\frac{1}{25,000}} = \left(\frac{1}{6}\right)\left(\frac{1}{500}\right) = \frac{1}{3000}$$

2. The answer is B. The presence of many affected persons in multiple generations of this family suggests autosomal dominant inheritance. Male-to-male transmission of the condition in several generations would rule out both X-linked and mitochondrial disorders. Each affected person has an affected parent and there are multiple affected persons in several successive generations.

3. The answer is B. The incidence of the disease among blacks, 1/400, can be used to determine q and p:

$$q = \sqrt{\frac{1}{400}} = 1/20 = 0.05 \text{ and } p = 1-q = 1-0.05 = 0.95$$

The frequency of heterozygotes is then calculated using Hardy-Weinberg Law:

$$Frequency = 2\,pq = 2(0.95)(0.05) = 0.095 \text{ or } \sim 1 \text{ in } 11$$

4. The answer is A. The pedigree demonstrates that the age of disease onset decreases in each successive generation, suggesting anticipation. This phenomenon occurs in diseases caused by expansion of a trinucleotide repeat in the disease-causing gene. In such disorders as Fragile X syndrome and Huntington disease, the disease occurs with greater severity or earlier age of onset in persons once a threshold length of the trinucleotide repeat tract in the gene is exceeded.

5. The answer is D. Neurofibromatosis is representative of many genetic diseases that exhibit pronounced variability of expression in individuals in families. This phenomenon is not understood at the molecular level but may involve the influence of environment

or other genes. Anticipation would have been a reasonable answer if the family history showed affected persons in multiple generations.

6. The answer is C. Several features of this pedigree suggest an X-linked mode of inheritance. All the affected persons are male, the disease tends to skip a generation, and there is no male-to-male transmission evident. Individuals III-2 and III-5, daughters of an affected father, are unaffected. This establishes that the condition is not dominant but recessive.

7. The answer is E. Individual II-7 is the mother of an affected son, so barring the occurrence of a new mutation in his case, she is an obligate carrier, ie, heterozygous for the condition.

8. The answer is B. The clinical symptoms in this case strongly suggest a mitochondrial disorder affecting both neurologic and musculoskeletal functions. In such cases, no male-to-male transmission is possible because the mother's ovum provides all the cytoplasmic components, including the mitochondria, for the fertilized egg. A pedigree for this family would resemble the inheritance pattern of an X-linked disorder with the likelihood of variable expression.

9. The answer is D. These syndromes involve reciprocal imprinting of one or more genes in a region of the long arm of chromosome 15. Although both diseases manifest neurologic impairment, Angelman syndrome is by far the more severe of the two disorders.

CHAPTER 14
CELLULAR SIGNALING
AND CANCER BIOLOGY

I. General Principles of Cellular Signaling

A. **Signaling** is a process by which **information** received from one cell is converted into a **response** by another cell.

1. The **signaling cell** produces and ordinarily secretes a **ligand,** which travels to the target cell.

2. **Receptors** displayed by the **target cell** bind the ligand and **transduce** the signal into a second messenger or a series of biochemical events that mediate the response by the target cell.

   a. The **second messengers** produced in response to these changes include **cyclic AMP, diacylglycerol (DAG), inositol trisphosphate (IP$_3$), and calcium.**

   b. Intermediate steps in the response involve alterations in the activity of enzymes, eg, **kinases** that phosphorylate cellular substrates.

   c. The ultimate physiologic response of the cell often involves **changes in gene expression.**

3. The signaling cell may be located far away from or adjacent to the target cell; in some cases, a cell signals itself.

B. The events set in motion by ligand binding to the receptor represent the **signaling pathway.**

1. In many cases, the signal is **transferred** or internalized from outside to inside the cell via the action of the receptor.

2. The signal may be **amplified;** in other words, activation of an enzyme at one step leads to multiple events as the signal is propagated through the pathway.

3. The steps of the signaling pathway also **distribute** the signal to several response pathways in the cell (eg, glucagon action to increase glucose production by the liver is mediated by increases in both gluconeogenesis and glycogenolysis).

4. The signal is often **modulated by integration** with input from other signaling pathways (eg, the insulin/glucagon ratio is important for balancing carbohydrate metabolism in the liver).

C. There is a variety of signaling modes used for intercellular communication.

1. In **endocrine** signaling, a **hormone** moves from a **gland** to a distant target cell by **direct secretion** into the bloodstream.

2. During neurotransmission, the **axon** of one neuron sends a chemical signal (a **neurotransmitter**) to receptors on the **dendrite** of another neuron across the **synaptic cleft.**

a. An important example of such neurotransmission via synaptic signaling is represented by the action of **acetylcholine** in cholinergic neurons.

    b. A closely related mechanism operates in neuroendocrine secretions (eg, arginine vasopressin and oxytocin release from hypothalamic neurons whose axons terminate at blood vessels within the posterior pituitary).

3. Paracrine signaling involves diffusion of a ligand from one cell to another **locally** within tissues.

    a. This signaling mode is commonly used by solid-tumor **growth factors,** such as epidermal growth factor (**EGF**) and the insulin-like growth factors **IGF-I** and **IGF-II,** for self-stimulation of continued cell division.

    b. Signaling by **nitric oxide** to regulate relaxation of smooth muscle cells also operates by the paracrine mode.

4. Juxtacrine signaling is a special case of the paracrine mode in which a growth factor **displayed on the surface of one cell** binds to a receptor on a neighboring cell.

5. Autocrine signaling occurs when a growth factor produced by a cell binds to a receptor on that **same cell.**

    a. This signaling mode is frequently used as a strategy for growth **self-stimulation.**

    b. Inappropriate **autocrine growth loops** resulting in uncontrolled cell growth are observed in many **cancer cells,** often involving EGF or IGF-II.

II. Signaling by G Protein-Coupled Receptors

A. Some cell-surface receptors having a characteristic structure with **seven membrane-spanning domains** couple to **heterotrimeric G proteins** for purposes of signal transduction (Figure 14–1).

B. Signaling is initiated by binding of an extracellular ligand, which produces a **conformational change** in the receptor that allows it to bind to the heterotrimeric G protein.

1. Binding of the ligand-receptor complex initiates a conformational change in the **G protein** that allows **exchange of GDP for GTP** on its α subunit, leading to **dissociation of the α and $\beta\gamma$** subunits of the G protein (Figure 14–1).

    a. The heterotrimeric G proteins reside in the plasma membrane and are composed of α, β, and γ subunits.

    b. Interactions between the α subunit and the $\beta\gamma$ complex are regulated by binding of ligand-receptor complex within the plane of the membrane.

2. Both α and $\beta\gamma$ can **bind to and regulate the activity of various effectors** of signaling.

    a. Regulation of the effectors may either be positive (activation) or negative (inhibition).

    b. Examples of some effectors are **adenylate cyclase, ion channels,** and **phospholipase C (PLC),** each of which regulates a different response pathway within the cell.

3. The signal is shut off by the action of the **intrinsic GTPase activity** of the α subunit, which catalyzes hydrolysis of the bound GTP to GDP and P_i.

    a. GTP hydrolysis to GDP + P_i causes a conformational change in the α subunit that returns it to the **inactive state.**

    b. The inactive α subunit dissociates from the effector and then binds a $\beta\gamma$ complex to be ready to undergo a new round of activation.

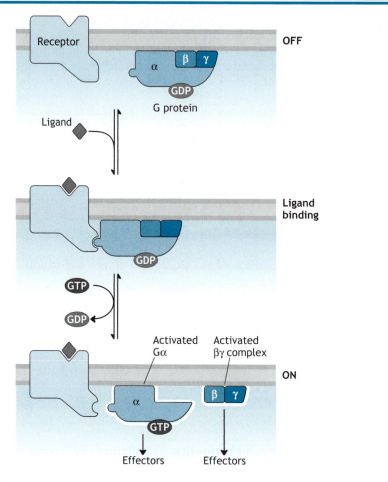

Figure 14–1. Signaling via G protein-coupled receptors. Ligand binding to its cell-surface receptor initiates interaction of the receptor with the heterotrimeric G protein for which it is specific. A conformational change in the G protein brought about by binding of the ligand-receptor complex promotes exchange of GDP for GTP. The activated G_α-GTP dissociates from the $G_{\beta\gamma}$ complex and both can interact with effectors, which carry on the signal to the mechanism that implements the cellular response.

4. The family of heterotrimeric G proteins that mediate signals from various cell-surface receptors is large (Table 14–1).
   a. Each G protein can be activated by one or more types of receptors.
   b. The G proteins are specific for a more limited range of effectors, which may be activated or inhibited by binding of the G protein α subunit.
   c. G protein signaling specificity and ligand binding functions reside in the α subunit; the βγ subunits for all members of the family are derived from a common pool in each cell.

Table 14–1. Heterotrimeric G proteins.

α Subunit	Effector and Directional Change
G_s (stimulatory)	↑ Adenylate cyclase and Ca^{2+} channels
G_i (inhibitory)	↓ Adenylate cyclase
G_o	↓ Ca^{2+} channels
G_q	↑ Phospholipase C
G_{11}, G_{14}, $G_{15/16}$	↑ Phospholipase C
G_t (transducin)	↑ Cyclic GMP phosphodiesterase
G_{Olf} (Olfactory receptors)	↑ Adenylate cyclase

    C. Second messengers mediate activation of signaling pathways downstream of the G protein–coupled receptors.
       1. Adenylate cyclase regulates synthesis of **cyclic AMP,** a second messenger that carries the signal through various cellular pathways to end points of response within the cell (Figure 14–2).
         a. Activation of **adenylate cyclase** stimulates synthesis of cyclic AMP from ATP.
         b. Cyclic AMP binds to protein kinase A (PKA), which dissociates into the active catalytic subunits and the regulatory subunits.
         c. The free **catalytic subunits** of PKA **phosphorylate cellular substrates** to change their activities and implement the cellular response.
           (1) Phosphorylation of cellular substrates by PKA occurs on serine and threonine residues.
           (2) Some examples of PKA substrates are **glycogen synthase** (see Chapter 6) and **CREB** (see Chapter 12).
         d. In order to shut off the signal mediated by elevated cellular cyclic AMP, **phosphodiesterase** hydrolyzes cyclic AMP to AMP (Figure 14–2).

METHYLXANTHINES

- *Several methylxanthines produced by plants are **inhibitors of cyclic AMP phosphodiesterase** and thereby produce an elevation in cyclic AMP levels in cells throughout the body.*
- *Effects of the methylxanthines are particularly pronounced in **cardiovascular and nervous systems** as increased heart rate, smooth muscle relaxation, and heightened activity of some neurons (perception of enhanced alertness).*
- *Methylxanthines include caffeine (present in coffee and tea), theobromine (tea and chocolate), and theophylline (tea).*
- *Due to its ability to stimulate relaxation of bronchial smooth muscle, **theophylline** is useful for treating the bronchoconstriction of asthma.*

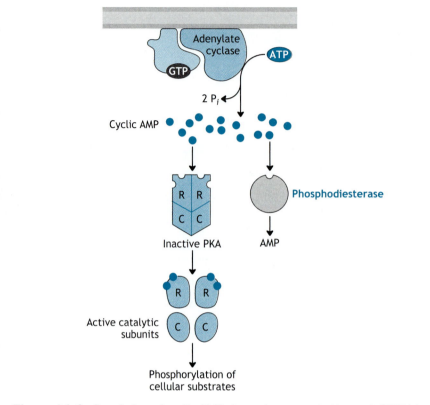

Figure 14–2. Regulation of cyclic AMP-dependent protein kinase A (PKA) by cyclic AMP. Activation of adenylate cyclase by binding of $G_{\alpha S}$-GTP amplifies the signal by synthesis of many molecules of cyclic AMP. Cyclic AMP binding to PKA causes dissociation of the regulatory subunits from the catalytic subunits, which carry on the signal. Phosphodiesterase regulates the concentration of cyclic AMP by catalyzing its hydrolysis to AMP, which shuts off the signal.

SOME BACTERIAL TOXINS INTERFERE WITH G PROTEIN FUNCTION

- *The toxins produced by* Bordetella pertussis *(**pertussis toxin**) and* Vibrio cholerae *(**cholera toxin**) modify functions of certain G proteins to produce their pathological effects.*
- *Cholera toxin is an enzyme that covalently modifies the α subunit of G_s by **ADP ribosylation,** which causes G_s to be in a persistently activated state.*
- *A consequent increased cyclic AMP concentration in the intestinal epithelial cells causes a large efflux of Cl⁻ and water into the gut producing the severe and potentially fatal diarrhea associated with cholera.*
- *Similarly, ADP ribosylation of the α subunit of G_i caused by pertussis toxin in the epithelial cells lining the airways of patients having a B pertussis infection inhibits exchange of GDP for GTP that ordinarily activates G_i.*
- *The resulting failure of G_i to inhibit adenylate cyclase increases cyclic AMP in airway cells and leads to fluid imbalance and the severe, life-threatening congestion of whooping cough.*

2. **Calcium and DAG are second messengers** that mediate some responses initiated by signaling from G protein-coupled receptors (Figure 14–3).
   a. Activation of **PLC** by binding of a G protein α subunit activates the enzyme.
   b. **PLC hydrolyzes** a membrane-bound inositol phospholipid, **phosphatidyl-inositol 4,5-bisphosphate (PIP$_2$)**, into the active products **IP$_3$** and **DAG**.
   c. **DAG** forms a **binding site for protein kinase C (PKC)** and thereby recruits it to the plasma membrane, which partially **activates** the enzyme.
   d. **IP$_3$ binds to the endoplasmic reticulum to release Ca^{2+}** stores.
   e. **Ca^{2+} binds to PKC and further activates it.**
   f. PKC **phosphorylates** multiple substrates to **alter gene expression** in the cell.

TUMOR-PROMOTING PHORBOL ESTERS

• *Extracts from the croton plant (croton oil) are not themselves carcinogenic but enhance tumor formation if administered after initial exposure to a carcinogen.*

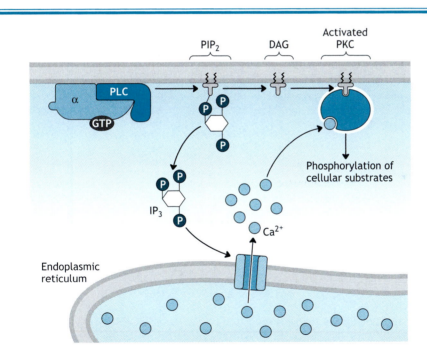

Figure 14–3. Signaling through protein kinase C (PKC). Activated phospholipase C cleaves the inositol phospholipid PIP$_2$ to form both soluble (IP$_3$) and membrane-associated (DAG) second messengers. DAG recruits PKC to the membrane, where binding of calcium ions to PKC fully activates it. To accomplish this, IP$_3$ promotes a transient increase of intracellular Ca^{2+} concentration by binding to a receptor on the endoplasmic reticulum, which opens a channel allowing release of stored calcium ions. PIP$_2$, phosphatidylinositol 4,5-bisphosphate; DAG, diacylglycerol; PLC, phospholipase C; IP$_3$, inositol trisphosphate.

- The active agents in croton oil are **phorbol esters**, specifically 12-O-tetradecanoyl phorbol-13-acetate (TPA) or phorbol myristate acetate (PMA), which are **structural analogs of DAG.**
- Both TPA and PMA enhance carcinogenesis by binding to the DAG binding site and activating PKC, which bypasses normal cell cycle regulation and **stimulates cell division** to produce its "tumor promoting" effect.

III. Receptor Tyrosine Kinases

A. Some cell-surface receptors transduce their signals by means of a kinase cascade initiated by their **protein tyrosine kinase** activity (Figure 14–4).

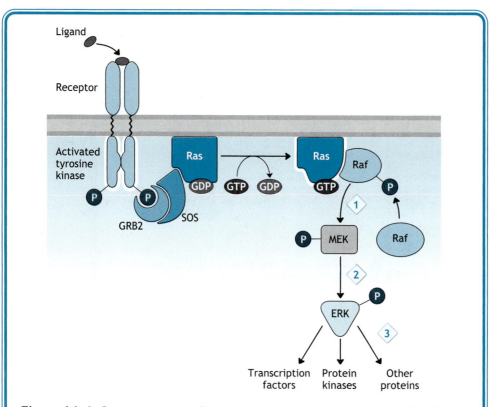

Figure 14–4. Receptor tyrosine kinase signaling mediated by the Ras-MEK-ERK pathway. Binding of a growth factor (ligand) to its cell-surface receptor promotes dimerization of the receptor with subsequent autophosphorylation mediated by activation of the intrinsic tyrosine kinase of the receptor's cytoplasmic domain. Docking of the adaptor GRB2-SOS complex promotes activation of Ras by GDP-GTP exchange. Ras recruits the first serine/threonine kinase of the signaling pathway, Raf. Raf then phosphorylates itself as well as the downstream kinase (MEK), which in turn phosphorylates ERK (also called MAP kinase). Activated ERK is capable of distributing the signal by phosphorylation of multiple substrates leading to the cell's pleiotropic response to the growth factor. Reactions of the kinase cascade are denoted by the numbers in *diamonds*.

1. These signal transducers have a large extracellular domain with its ligand-binding site, a single transmembrane domain and an intracellular domain with intrinsic tyrosine kinase activity.
2. **Ligand binding** to the receptor's extracellular domain activates signaling by causing the **receptors to form dimers** and cross-phosphorylate (**autophosphorylate**) their intracellular domains on tyrosine sites.

B. The signaling pathway downstream of the activated receptors is composed of a series of kinases, a **kinase cascade.**
   1. The **phosphotyrosine sites** on the receptor act as **docking points for adaptors and effectors,** which couple the signal to the kinase cascade.
   2. One of the major adaptors is the **GRB2-SOS** complex, which upon docking to the phosphorylated receptor, binds the **small G protein Ras** and activates it by **GDP-GTP exchange** in a manner analogous to the heterotrimeric G proteins.
   3. **Activated Ras recruits** the first kinase in the cascade, **Raf-1, to the plasma membrane,** where it becomes active.
   4. The signal is then transferred from one kinase to the other by sequential phosphorylation and activation, ie, the kinase cascade.
   5. The signal ultimately is sent into the nucleus, where **transcription factors** such as Elk-1 are activated by phosphorylation.

CLINICAL APPLICATIONS OF MONOCLONAL ANTIBODIES THAT TARGET LIGANDS AND RECEPTORS

- *By 2005, 18 monoclonal antibodies had been approved for treatment of several diseases, especially for various* **cancers** *as well as* **infectious and inflammatory conditions,** *with many more under development.*
- *Some of these agents are targeted to ligands or their receptors, and they work by preventing binding and subsequent signal transduction, as illustrated in the following examples.*
  - **HER2,** *a member of the EGF receptor family, drives growth of breast cancers that overexpress the receptor. Trastuzumab, which binds HER2 and prevents receptor activation, has been shown to be effective in reducing tumor growth and metastasis in such cases.*
  - **Interleukin-2 (IL-2)** *signaling is important in the immune response that can lead to rejection in solid organ transplantation. Basiliximab binds the α subunit of the IL-2 receptor to prevent IL-2 binding and provide an immunosuppressive effect to inhibit renal transplant rejection.*
  - **Tumor necrosis factor-α (TNF-α)** *is a critical mediator of inflammation in autoimmune diseases like Crohn's disease and rheumatoid arthritis. Infliximab binds TNF-α and prevents its binding to the TNF receptor for treatment of these diseases.*
  - *Many cancers depend on* **vascular endothelial growth factor (VEGF)** *for formation of a blood supply to allow tumor growth and metastasis. Bevacizumab binds VEGF, which prevents its binding to the VEGF receptor and thereby inhibits tumor vascularization (angiogenesis) in combination therapy with 5-fluorouracil for treatment of metastatic cancers, particularly colorectal cancer.*

IV. The Nuclear Receptor Superfamily

A. Many hormones diffuse into the cell and initiate signaling by binding to **soluble intracellular receptors** that act as **transcription factors.**
   1. This mechanism is used by **steroid hormones** (Table 14–2), thyroid hormone, vitamin D_3, and retinoic acid.
      a. These ligands for the nuclear receptor superfamily are capable of dissolving in water at low concentrations but are mainly **lipophilic,** capable of passing through the lipid bilayer into the cell by diffusion.

Table 14–2. Ligands of the nuclear receptor superfamily.

Hormone or Ligand Family Name	Major Ligands in Humans
Glucocorticoids	Cortisol
Mineralocorticoids	Aldosterone
Progestins	Progesterone
Estrogens	Estradiol Estriol Estrone
Androgens	Testosterone Dihydrotestosterone (DHT) Dehydroepiandrosterone (DHEA)
Vitamin D compounds	1,25-Dihydroxycholecalciferol or 1,25-Dihydroxy vitamin D_3
Retinoids (vitamin A compounds)	All-*trans* retinoic acid
Thyroid hormones	Thyroxine (T_4) Triiodothyronine (T_3)

    b. Some of these molecules **require metabolism** or modification to be able to bind their receptors.

        (1) **Dihydrotestosterone** is the preferred (high affinity) ligand for the **androgen receptor** and is formed by **reduction of testosterone** catalyzed by the enzyme steroid 5α-reductase.

        (2) The form of thyroid hormone active in binding its receptor is **triiodothyronine** (T_3) rather than thyroxine (T_4).

  2. The receptors may be located in the nucleus or cytoplasm of the cell, but they are collectively called the "**nuclear receptor superfamily**" because the nucleus is their main site of action.

  3. The receptors in this family have a similar overall structure with a **ligand-binding domain** specific for the hormone or vitamin, a **DNA-binding domain,** and a **variable domain** that differs among the receptors.

B. Binding of ligand activates the receptor so that it can bind specific DNA sequences in regulatory regions of target genes that have **hormone-response elements (HREs)** (Figure 14–5).

  1. After formation of the initial **ligand-receptor complex,** other partner proteins are recruited that complete the active complex.

  2. Binding of a **co-activator** confers on the complex the ability to **activate transcription** when it binds to the target gene.

  3. Conversely, transcription of a target gene may be **inhibited** by binding of a complex formed when a **co-repressor** binds to the ligand-receptor.

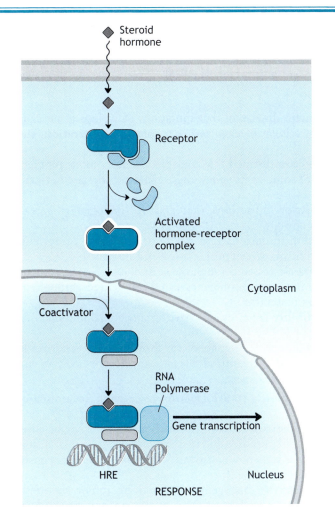

Figure 14–5. Regulation of gene transcription by members of the nuclear receptor superfamily. Binding of a steroid hormone to its receptor promotes a conformational change that causes dissociation of proteins that associate with the inactivated receptor, including several heat shock proteins. In this example, the receptor is localized in the cytoplasm in its inactive state. In such a case, the activated hormone-receptor complex undergoes a conformational change that exposes a nuclear localization signal. Within the nucleus, the receptor binds a coactivator protein and the complete complex mediates transcriptional activation of target genes having the appropriate hormone-response element (HRE).

DISORDERS OF ANDROGEN ACTION PRODUCE FEMINIZATION IN MALES

- **Steroid 5α-reductase deficiency** is an autosomal recessive disorder that causes decreased conversion of testosterone to dihydrotestosterone and decreased androgen action that is particularly critical during sexual development.

- *External genitalia of men deficient in steroid 5α-reductase are female in character rather than male.*
- *Several inherited disorders that produce **defective androgen receptors** (androgen resistance) also cause disruption of sexual development that may culminate in infertility or testicular feminization.*
- ***Testicular feminization** is characterized by expression of a female external phenotype despite a normal blood level of testosterone and standard **male karyotype** (46,XY).*

V. Overview of Cancer Biology

A. **Cancer** is considered a **genetic disease** in that mutations of various genes cause disease by **dysregulation** of cellular mechanisms that control **proliferation, survival, and death.**

   1. Once a cell has become **"transformed,"** ie, capable of autonomous proliferation through **mutation** of some of its genes, these characteristics are **heritable from cell to cell.**

   2. **Dominant, gain-of-function mutations that activate oncogenes** confer a rapid-growth phenotype on cells.

   3. **Recessive, loss-of-function mutations that delete or inactivate tumor suppressor genes** alleviate controls on cell proliferation and survival.

   4. Activated oncogenes are rarely passed through the germline.

   5. Mutated, inactivated tumor suppressor genes can be inherited through the germline from one person to another.

      a. These **cancer susceptibility** genes usually have an autosomal dominant expression pattern.

      b. Examples of such conditions are the genes for **familial colorectal cancer** (eg, *HNPCC* or *APC*) and the **familial breast cancer** genes *BRCA1* and *BRCA2.*

B. Development of cancer or **neoplastic transformation** requires an **accumulation of mutations in the same cell.**

   1. The **first mutation** in a tumor suppressor gene such as *BRCA1* may be either inherited via the germline or **sporadic** (due to a random event in that person) and then the normal allele is somehow inactivated (see **loss of heterozygosity** below).

   2. Multiple mutations that activate oncogenes or inactivate tumor suppressor genes accumulate due to progressive **loss of DNA repair mechanisms and cell cycle control.**

   3. An important example of how a progression of somatic mutations leads to cancer is in hereditary colorectal cancer (Figure 14–6).

VI. Oncogenes and Tumor Suppressor Genes

A. **Oncogene** activation by overexpression, mutation, or chromosomal rearrangement can lead to rapid proliferation of cells and cancer.

   1. Oncogenes are the mutant, out-of-control versions of normal cellular genes, the **proto-oncogenes,** which regulate a variety of critical cellular processes such as signaling, cell cycle control, and transcription.

   2. The mutations that have converted the proto-oncogenes to their oncogene forms are **gain-of-function** or **activating mutations.**

RAS MUTATIONS OCCUR IN MANY HUMAN CANCERS

- *Over 30% of all human cancers have activating mutations of the gene encoding the small G protein Ras.*

CLINICAL CORRELATION

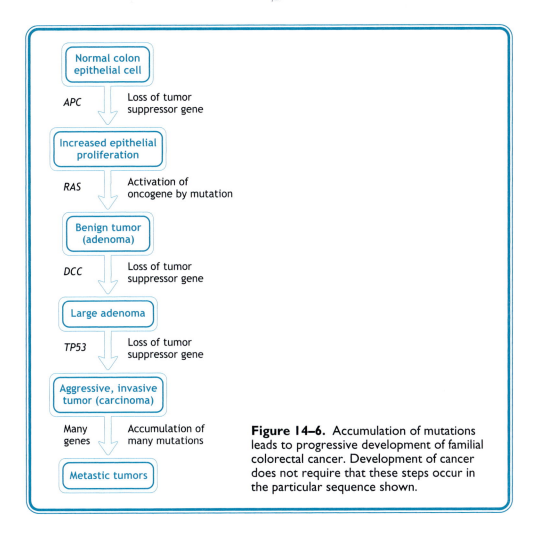

Figure 14–6. Accumulation of mutations leads to progressive development of familial colorectal cancer. Development of cancer does not require that these steps occur in the particular sequence shown.

- *Several **missense mutations** (ie, at codons 12, 13, or 61) render the mutant protein incapable of hydrolyzing bound GTP to GDP.*
- *These mutant forms of Ras thus persist in the **ON state,** which provides continuous activation of the kinase cascade downstream of Ras and stimulates the cell to keep dividing even in the absence of appropriate signals from cell-surface receptors.*

   3. Tumor viruses carry activated versions of important cellular genes that regulate cell cycle and transcription.
     a. The virus that causes Kaposi's sarcoma, **Kaposi's sarcoma–associated herpesvirus**, induces transformation of infected cells by up-regulating expression of the cellular form of the **Kit oncogene,** among others.
     b. Human papillomavirus (HPV) causes a variety of epithelial cancers, especially of the alimentary canal and the cervix, by means of two associated **oncogenes, E6 and E7.**
   4. Overexpression or deregulated expression of cell cycle-dependent transcription factors such as **Myc** and **Fos** may stimulate continued cell division.

5. Activation of an oncogene may occur by **chromosomal rearrangement** creating a dysregulated **fusion protein.**

THE PHILADELPHIA CHROMOSOME IN CHRONIC MYELOGENOUS LEUKEMIA

- *Cytogenetic analysis of patients with chronic myelogenous leukemia (CML) reveals an unusual translocation between chromosomes 9 and 22 termed the "Philadelphia chromosome."*
- *The translocation moves the c-ABL gene that encodes a tyrosine kinase from chromosome 9 to the breakpoint cluster region (BCR) of chromosome 22.*
- *The resultant gene, **BCR-ABL,** encodes a **constitutively active kinase** that stimulates cell division and leads to the transformed phenotype of the cells.*
- *Patients with CML experience weakness, fatigue, excessive sweating, low-grade fever, enlarged spleen, and elevated WBC count.*
- *Imatinib, a drug that inhibits the kinase activity of the Bcr-Abl fusion protein, has been successfully used for treatment of CML.*

    B. Loss or inactivation of tumor suppressor genes may lead to cancer.

        1. Tumor suppressors are genes that encode a diverse array of proteins that **control cellular growth and death.**

        2. Loss or mutation that inactivates one copy of the gene can be tolerated because there is no functional deficit in the heterozygous condition.

        3. Loss of heterozygosity (LOH) that deletes the only available functional copy of the gene can contribute to unregulated proliferation of those cells (Figure 14–7).

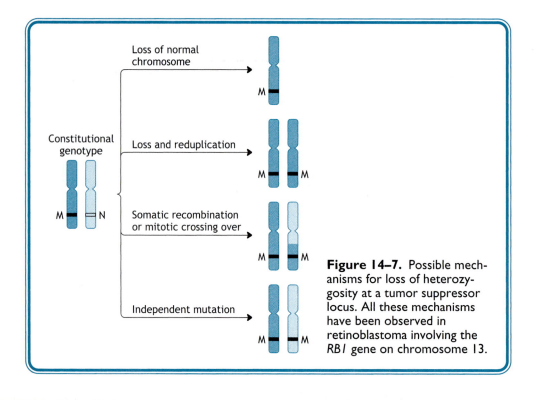

Figure 14–7. Possible mechanisms for loss of heterozygosity at a tumor suppressor locus. All these mechanisms have been observed in retinoblastoma involving the *RB1* gene on chromosome 13.

LOH IN RETINOBLASTOMA

- *Retinoblastoma produces childhood neoplasms arising from neural precursor cells of the retina (**retinoblasts**) at an incidence of 1 in 20,000 live births.*
- *The biochemical defect is mutation or loss of the **tumor suppressor gene, RB1,** encoding the protein pRb.*
  - *pRb binds to and inactivates members of the E2F transcription complex, which normally prevents cells from entering S phase of the cell cycle.*
  - *Loss of E2F regulation by pRb impairs cell cycle control, and unregulated proliferation (**clonal expansion**) may lead to a **tumor** derived from that cell.*
- *Most cases are inherited and multiple tumors arise bilaterally in heterozygotes when the normal RB1 allele undergoes mutation or loss due to LOH.*
- *Retinoblastoma shows an **apparently autosomal dominant phenotype** due to the high probability of LOH during the ~10^6 cell divisions of retinoblasts and despite the recessive nature at the cellular level.*

    4. *TP53* is an important tumor suppressor gene that encodes the **p53** transcription factor that is up-regulated when the cellular **DNA is damaged.**

        a. High levels of p53 up-regulate transcription of the *WAF1/CIP1* gene, whose protein product, p21, blocks entry into S phase of the cell cycle by a mechanism called **checkpoint control.**

        b. *TP53* is the most commonly mutated gene in human cancer, occurring in over 50% of tumors examined.

LI-FRAUMENI SYNDROME

- *Patients with Li-Fraumeni syndrome have **increased susceptibility to a variety of cancers,** including bone and soft-tissue sarcomas, breast tumors, brain cancers, leukemia, and adrenocortical carcinoma, all arising at an **early age** (often before 30 years).*
- *The biochemical defect in families exhibiting this syndrome is a **loss-of-function mutation** of the tumor suppressor gene, **TP53,** encoding p53.*
- *The incidence of Li-Fraumeni syndrome has not been calculated because it is so rare.*
- *Inheritance is **apparently autosomal dominant** with high penetrance but with **variable expression** (family members may have a wide range of tumor types and ages of onset).*

VII. Apoptosis

    A. Apoptosis, or programmed cell death, is a complex, highly regulated process by which a cell self-destructs in an organized manner.

        1. The mechanism of death in apoptosis contrasts with that occurring when a cell breaks open or lyses producing a **necrosis.**

        2. Necrosis allows the contents of the cell to spill over the local area, causing an **inflammatory response** that leads to damage to nearby cells within the tissue.

        3. By contrast, cells undergoing apoptosis do not lyse, so there is no associated inflammatory response.

    B. Major changes that occur in the cell during apoptosis include the following:

        1. Chromatin condensation.

        2. Disintegration of the nuclear envelope.

        3. Fragmentation of DNA between the nucleosomes.

        4. Blebbing of the cell membrane.

        5. Recruitment of macrophages, which ultimately engulf the dead cells.

    C. Both extrinsic and intrinsic pathways can lead to programmed cell death (Figure 14–8).

1. The **extrinsic pathway** involves response to an **external signal.**

   a. The external signal of **death ligands,** such as FasL and tumor necrosis factor–related apoptosis-inducing ligand (TRAIL), is transduced by cell-surface **death receptors,** such as FADD.

   b. Activation of an array of **proteases** called caspases (the **caspase cascade**) mediates the response within the cell, which involves initiator caspases that cleave and activate effector caspases.

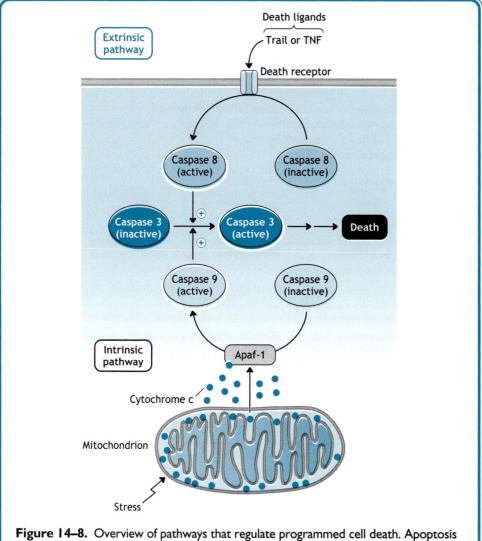

Figure 14–8. Overview of pathways that regulate programmed cell death. Apoptosis may occur in response to signaling through either the extrinsic pathway or the intrinsic pathway. In each case, proteolytic cleavage activates an initiator caspase, caspase 8 or 9, either of which can cleave an effector caspase such as caspase 3. Apaf-1 is part of a large complex called the apoptosome that mediates the intrinsic pathway. Binding of an extracellular death ligand to its cell-surface receptor activates the extrinsic pathway.

c. Effector caspases in turn degrade key cellular proteins and activate an **endonuclease** that digests the DNA.
  2. The **intrinsic pathway** responds to **stress,** usually resulting in the cell's **inability to repair extensive DNA damage,** sparking a decision to commit suicide.
      a. Activation of **pro-apoptotic (death-causing) factors** may occur in response to the DNA damage, which causes **increased mitochondrial permeability.**
      b. **Leakage of cytochrome c,** among other proteins, from the intermembrane space of the mitochondria causes activation of the caspase cascade.

CLINICAL PROBLEMS

A 19-year-old woman has been referred to an endocrinologist by her gynecologist because of delay in the initiation of her menstrual periods. Physical examination reveals underdeveloped breasts, an enlarged clitoris (rudimentary penis), and the presence of small masses within the labia majora. Blood testosterone is in the normal range for males and a chromosome spread indicates a karyotype of 46,XY.

1. This patient most likely has a defect in signaling through a pathway involving which of the following?

  A. Cyclic AMP–dependent protein kinase (PKA)

  B. Protein kinase C (PKC)

  C. A cell-surface tyrosine kinase receptor

  D. A nuclear receptor

  E. A heterotrimeric G protein

In order for a solid tumor to grow beyond a certain size, it must develop a blood supply by elaborating factors such as vascular endothelial growth factor (VEGF). VEGF secreted by the tumor cells diffuses to nearby endothelial cells, which respond by dividing and migrating toward the tumor to eventually develop into blood vessels and vascularize the tumor.

2. Which of the following modes of intercellular signaling is operative in the case of VEGF?

  A. Endocrine

  B. Paracrine

  C. Autocrine

  D. Juxtacrine

  E. Synaptic

Many of the drugs used in the treatment of hypertension and cardiovascular disease are designed to interfere with the action of cell-surface receptors that couple to heterotrimeric G proteins.

3. In order for these drugs to operate in a specific manner so that cellular responses to only one type of receptor are affected, the drug would need to be targeted toward which element of the pathway?

A. The ligand binding site of the receptor

B. The βγ complex of the G protein

C. The α subunit of the G protein

D. Adenylate cyclase

E. Phospholipase C

It is estimated that mutations of *RAS* occur in over 30% of human cancers. In most of these cases, the mutations interfere with the intrinsic GTPase activity of Ras so that the protein becomes constitutively or continuously active, irrespective of whether growth factors are present.

4. Constitutively activated Ras has become insensitive to which of the following elements of the growth factor signaling pathway?

A. Raf-1

B. MEK

C. MAP kinase

D. Ras-GAP

E. Elk-1

Patients with retinoblastoma suffer from a high incidence of tumors arising from clonal outgrowth of some retinal precursor cells due to mutation of the tumor suppressor gene *RB1*. Analysis of cells from these tumors indicates that both copies of the *RB1* gene are mutated or lost, whereas the surrounding retinal cells have at least one functional *RB1* allele.

5. Which of the following terms best describes the genetic phenomenon that leads to tumor development in retinoblastoma patients?

A. Loss of imprinting

B. Deregulated expression

C. Incomplete penetrance

D. Gain of function

E. Loss of heterozygosity

Osteosarcoma has recently been diagnosed in a 12-year-old girl. Family history indicates that her paternal aunt died of breast cancer at age 29 after having survived treatment for an adrenocortical carcinoma. An uncle died of a brain tumor at age 38 and the patient's father, age 35, has leukemia.

6. An analysis of this patient's DNA would most likely reveal a mutation in which of the following genes?

A. *RB1*

B. *RAS*

C. *TP53*

D. *c-ABL*

E. *PKC*

ANSWERS

1. The answer is D. The patient's ambiguous secondary sex characteristics and lack of menstrual activity suggest the possibility of an androgen resistance syndrome. The male karyotype and blood testosterone levels confirm this. This clinical condition might have arisen as a result of steroid 5α-reductase deficiency or inherited defects in the androgen receptor (testicular feminization).

2. The answer is B. Paracrine signaling involves diffusion of a substance locally from one cell to another via the interstitial space rather than through blood vessels. Endocrine signaling would require that VEGF travel through the blood to reach the endothelial target cells. Autocrine signaling requires that the same cell both send the signal and respond to it. Juxtacrine signaling requires that the VEGF be displayed from the surface of one cell and bound by a receptor on another. Synaptic signaling is reserved for neurons. None of these other signaling modes fit the description for the mechanism of action of VEGF.

3. The answer is A. Most of the drugs that target specific types of G protein-coupled receptors are either agonists that bind to the ligand-binding site and stimulate receptor activity or are antagonists that bind to the receptor and prevent ligand binding. The G protein α and βγ subunits, adenylate cyclase, and phospholipase C are all elements shared among many types of receptors.

4. The answer is D. In response to binding of a growth factor to its cell-surface receptor, the receptor forms a dimer that stimulates its intrinsic kinase activity to phosphorylate tyrosine residues on the cytoplasmic region. These phosphotyrosine sites allow docking of the adaptor complex GRB2-SOS, which binds and thereby activates Ras through GDP to GTP exchange. Constitutively activated Ras is unable to hydrolyze bound GTP and thus cannot respond to the binding of Ras-GAP. Raf-1, MEK, MAP kinase, and Elk-1 all are downstream elements of the signaling pathway that depend on the activity of Ras.

5. The answer is E. At the cellular level, the *RB1* gene is recessive because loss of function affecting both alleles must occur to produce disease. This patient has inherited a defective *RB1* allele from her father and is thus heterozygous at the *RB1* locus. Most of her retinal precursor cells have one functional *RB1* allele and those cells proliferate under normal growth restraints. However, these cells are susceptible to mutations affecting pRb function or an error leading to loss of the remaining functional *RB1* allele. These mutations occur by chance during cell division and lead to a tumor by clonal outgrowth. The process by which the sole functional allele is lost or mutated is referred to as loss of heterozygosity (LOH).

6. The answer is C. The occurrence of a variety of cancers at fairly early ages in this family, particularly the finding of osteosarcoma in such a young girl, suggests the possibility of an inherited disorder of a tumor suppressor gene. Since the tumors are not associated with the eye, *RB1* is unlikely as the cause. The spectrum of cancers in the family is consistent with the Li-Fraumeni syndrome, which involves inheritance of a loss-of-function mutant form of the tumor suppressor gene, *TP53*, encoding p53.

INDEX

Note: Page numbers followed by *f* or *t* indicate figures or tables, respectively.